Being Human

P.A. Abdo

Copyright © 2014 by P.A. Abdo.

Library of Congress Control Number:		2014911033
ISBN:	Hardcover	978-1-4633-8707-5
	Softcover	978-1-4633-8706-8
	eBook	978-1-4633-8705-1

All rights reserved. No part of this book may be reproduced or transmitted in any form or by any means, electronic or mechanical, including photocopying, recording, or by any information storage and retrieval system, without permission in writing from the copyright owner.

The views expressed in this work are solely those of the author and do not necessarily reflect the views of the publisher, and the publisher hereby disclaims any responsibility for them.

This book was printed in the United States of America.

Rev. date: 18/06/2014
Title: Being Human
Author: P.A. Abdo
p.a.abdo@beinghuman-thebook.com
Cover: Carlos Villaseñor Múzquiz
Translation: Jorge A. Salazar
English edits to Translation: Thomas Frederick Bartlett

Copyright © First edition Spanish: August 2013
Public Registry No. 03-2013-091312165400-01
Cover Design Registry No. 03-2013-091312213900-01
Copyright © First edition English: May 2014
Scripture taken from the King James Version of the Bible.

To order additional copies of this book, please contact:
Palibrio LLC
1663 Liberty Drive
Suite 200
Bloomington, IN 47403
Toll Free from the U.S.A 877.407.5847
Toll Free from Mexico 01.800.288.2243
Toll Free from Spain 900.866.949
From other International locations +1.812.671.9757
Fax: 01.812.355.1576
orders@palibrio.com

CONTENTS

INTRODUCTION 7

PRELUDE
    Living Things 11

FIRST PART
THE LAWS OF NATURE

CHAPTER I
    Historical references 27

CHAPTER II
    The laws that govern Nature 35

CHAPTER III
    Science 53

CHAPTER IV
    The laws affecting all living creatures 65
    The Soul and the Life 65

SECOND PART
LAWS GOVERNING
THE HUMAN BEING IN PARTICULAR

CHAPTER I
    The Spirit 79

CHAPTER II
    Instincts 95

CHAPTER III
    The Universal Conscience and the Human
    Conscience 99

CHAPTER IV
    Moral 105

CHAPTER V
    Ethics 119

CHAPTER VI
    Love 127

CHAPTER VII
    Homosexuality 135

CHAPTER VIII
    Human Rights 145

CONCLUSION 155

BIBLIOGRAFÍA 159

Dedication:

*To an extraordinary individual,
someone that changed my life
and who asked me to be
kept anonymous.*

Introduction

Why is there so much disorder in the world? What is the cause of war, violence, overpopulation…? What is the origin of passions such as hate, ambition, resentment, vengeance, envy and others? What causes obesity? Why do poverty, hunger, drought, deforestation, and pollution exist in the world, among other deplorable situations?

Many authors try to explain all these issues by expressing different opinions and premises –in many cases- antagonistic ones. Nevertheless, to this date, I have not found a text clearly answering in a satisfactory and informed manner the main causes of these issues.

This book intends to find an answer to all these questions from the perspective that there is no doubt humans created all imbalances in the world we live now.

This book shows the creative and destructive power source that lies inside each individual.

Is there a universal criterion –and an infallible one- that is able to explain the origin of all those evils?

Indeed there is. Through many years of observation, investigation and deductive consideration, the author has begun to understand that the world is designed to be in perfect balance and that it is human behavior what breaks it.

The thoughts expressed in this book not only explain the origin of such erratic behavior of all human beings, but it also brings canons of conduct that will improve our life quality, since they are universal ones that may be applied to all humanity with no distinction on race, creed or social status.

This book studies the human being as no one has seen it before: as a regulated creation. In fact, the human being –as part of the Nature- is not a being alien to the natural laws, which are invariable and eternal. Human actions and thoughts are governed by these norms, drifting from the popular belief that the human being freedom consists on doing whichever pleases every one.

Throughout the text of *BEING HUMAN* you will find a well-ordered display of arguments and deductions explaining the life of men and its sense, answering the most common questions about the effect of living. When you finish reading it, you will have, at your disposal, the knowledge to improve, in a significant manner, the quality of your life, even finding ways to increase your longevity.

BEING HUMAN is more than a book. It is an answer to the great question of why we are humans. It is a way to see the superior life that we have denied ourselves by choosing a culture that confuses materialism with comfort, sexuality with love, domination with success and image with what ought to be.

Access to the full life you have been waiting for is only a few pages away.

Eduardo José Abdo Cantu.

Prelude

Living Things

The Creator of the Universe has a specific purpose for all living things populating the Earth. This can be deduced from the plethora of evidence confirming this assertion, which I shall try to demonstrate during the course of my work.

Let us take as an example the creation of the massive luminous star we know as the Sun and that the Creator of the Universe made exclusively for terrestrial life.

Have you ever thought that living organisms populating Planet Earth are the only ones benefiting from the Sun in the whole universe?

The influence of the Sun only reaches out to Pluto. After such asteroid, previously considered as a Planet, there is no solar influence determining life.

All life existing in the Earth depends on the Sun. without it, no life would exist on the Earth. The size of the Sun and the distance between the Sun and our Planet are ideal so that terrestrial life can be developed in perfect equilibrium.

Do you think this is just luck?

The closest potentially inhabitable planet (most similar to Planet Earth), outside our Solar System, is situated in the binary system called Alfa Centaury orbiting the Alfa Centauri B star, 41.3 trillion kilometers (approximately 25.6 trillion miles) from our Sun.[1]

Dear reader, do you consider that a living organism from another Planet is able to travel such distance or even larger distances to come to Earth?

This fact prompts me to think that the Planet Earth inhabitants are fully isolated from any other intelligent life that may exist in the Universe.

Studies regarding extraterrestrial life are common now; nevertheless, to this date, there is no hard evidence to prove the existence of any form of life generated outside Planet Earth.[2] The existing hypotheses are just that, hypotheses.

Now, another obvious question comes to mind, do you think human beings are able to live in another place other than Planet Earth?

---

[1] BBC WORLD NEWS.- Closest-Ever Exoplanet Discovered
[2] Astrobiology – Wikipedia – The free encyclopedia.- Digital Version: http://en.wikipedia.org/wiki/Astrobiology

I really think that human beings are not able to live outside this Planet because the gravity force of the Earth is indispensable for the proper function of any organism;[3] besides, it also requires water, air and food produced in the Earth in order to survive. The amounts of these elements taken to a space expedition are small and their recycling limited.

This fact also makes me think that the reason the Creator placed living organisms on this Planet in such isolated conditions is because the Creator has a purpose for all living creatures, and it is the obligation of all human beings to look for such purpose.

This last assertion is accurate because all creation serves a purpose, even those that apparently have no logical explanation to a human.

On the other hand, talking about questions, why do you think all birds are oviparous, that is, lay eggs?

Taking into consideration that everything in Creation serves a purpose, a practical answer may be: if birds were viviparous, giving live birth, they would not be able to fly when pregnant because moving their young while flying would cause the bird to lose balance and fall to the earth, making it easy prey for predators.

---

[3] NASA: el organismo humano no es apto para viajar a Marte (The human organism is not apt to travel to Mars):
https://mx.finanzas.yahoo.com/noticias/nasa-el-organismo-humano-no-es-apto-para-021700145.html

But the Creator who designed the birds, wanted birds to produce a solid shell for their young to grow inside without being a bother; besides, the egg grows in the exact gravity center of the bird's body allowing it to keep its balance and fly free, thus, performing the roll assigned to it by Nature.

Another surprising fact from Creation showing equilibrium in Nature is also the percentage of men and women in our Planet; world population is composed of approximately fifty percent men and fifty percent women.[4]

This is not just luck either. There have been moments during war times that a country decreases its percentage of men[5] but at the next census such percentages magically equate[6] notwithstanding any planning made by humans to program the gender of their descendants.

This astonishing balance maintained over years is a result of the order of Nature and makes us conclude the following:

1st. There is a balance in Creation, even if we humans do not understand such balance.

---

[4] Total Population by Gender and Gender Ratio, by Country. Statistics Division, Department of Economic and Social Affairs.- *"World Population Prospects: The 2008 Revision"*.- Digital Edition:
http://www.geohive.com/earth/pop_gender.aspx
http://www.poodwaddle.com/clocks/worldclockes/

[5] The Book of the Year 1988.- Británica World Data.- *Encyclopædia Britannica*.- Ed. Encyclopædia Britannica, Inc.- Pag. 746.

[6] Book of the year 1998.- *Encyclopædia Britannica*.- Ed. Encyclopædia Britannica, Inc.- Pag. 756.

2nd. That this balance is not the product of accident nor that such balance maintains itself, by itself, but that this balance is governed by laws of mathematical precision.

3rd. Humans have been created to live in pairs, one man and one woman.

Besides the foregoing, we can also observe that all deeds, processes or phenomena occurring in Nature have a defined pattern of behavior.

Why do you think those events repeat invariably?

I think we owe this to the systematic respect for certain laws. We call them the laws of Nature, and their study is the purpose of this work.

These laws are invariable, and they are always governing the behavior of all Nature.

Now, I would like to make an intellectual exercise since the concerns I am planting in this prelude shall serve as a communion with the reader in the mental walk we are to take during the course of this book:

In Muir Woods national park located in the State of California, in the United States, there is a group of giant trees known as Giant Sequoias that measure up to 380 feet in height and 23 feet in diameter with roots reaching 114 feet deep.

How can a tree seek water by itself, then absorb it from the depth of its roots while selecting the minerals it needs, then conduct it fluently through its trunk together with the selected nutrients to feed even the last leaf of the tree at such incredible height and nourish its wide trunk all at the same time?

This process is due to an astounding feeding system making the tree function due to the information that the tree has incorporated into its genetic code.

This information is recorded in each tree of the same species, and it makes them work in the same manner. This information is one of the many laws governing all Nature, and we study them widely in the second chapter of this work.

These laws are identical for all species of the plant kingdom, and they regulate their metabolic functions, sap circulation, photosynthesis, nutrition system, development, reproduction and its defense methods.

This information is recorded in an indelible manner in the genetic code of all seeds of each plant species, and it is transmitted by inheritance making each species function identically to its predecessor having the same interior and exterior form.

Animals have a similar programming, and I would like the reader to ask him self the following question:

As I define obese as excess body fat, why are there are no obese animals in the Nature?

Whales, hippopotami, walruses, seals, etc., are not obese. There are no fat whales, hippopotamus, walruses or seals. The design of these animals as we observe them in the Nature is perfect.

I consider that the reason why we cannot find fat animals in Nature is because the laws governing their nutrition are genetically programmed for them to eat just what they need to live, and they do not naturally breach those laws. Animals do not exceed their nutrition limits unless they are captive because their natural equilibrium breaks.

But humans, through their own volition (free will) have the capacity to breach those laws governing their nourishment and exceed them when satisfying their hunger instinct, thus damaging their body as we can see more clearly in the chapter dedicated to the study of morals.

Humans are obligated to comply with what the laws of Nature order so that they do not damage themselves or their surroundings, and it compels them to do so in certain order.

Order is the purpose of the laws of Nature and of course, the purpose of the civil laws dictated by the human society so that they are congruent with the Nature, thereby making individuals able to live in peace and harmony.

Human beings are responsible for keeping Creation. Humans or obligated to do so for their own survival and the survival of all living things populating the Earth; nevertheless, the same human being is the one who has accelerated its destruction by disobeying those laws.

This is because humans have no knowledge of most of the laws regulating Nature, which laws we will study broadly in the second chapter of this book.

On the other hand, I would like to ask another question for my readers: why do all body organs (heart, kidneys, pancreas, etc.) in all humans beings have the same form, are located in the same place, and function identically in each one of them?

These questions go unnoticed for the majority of the population who ignore that their body is designed to obey the laws of Creation incorporated into their genetic code and do not know how to treat their own bodies.

Therefore, the purpose of this book is to try to transmit the way the laws of Nature regulate not only their bodies, but also the bodies of each living thing, managing the form and function of their organs such as digestion, metabolism, growth, etc.

In the same manner, instincts, behavior, the manner of finding food, the defense methods, and reproduction systems are identical en each member of the same species.

On the other hand, besides the laws mentioned in the previous paragraphs, humans have special qualities that distinguish them from all other living species.

The qualities or features exclusive to human beings are mainly: spoken language (as commonly defined), producing and understanding ideas, discerning between right and wrong, reflecting on their own selves, on their own sensations, on their own ideas, creating an idea and wanting to put that idea to work, and feeling the need for a Superior Entity.

Additionally, human beings have the exclusive faculty to seek their own longevity, take care of the environment, and produce their own science and technology to provide supplies for their own subsistence as well as for those animals and plants need.

These and other qualities will be studied when we talk about the spirit because this entity is exclusive for the human beings and all those attributes are spiritual qualities.

But what are the differences between the complicated concepts of soul and spirit, and which functions are performed by each one of those entities?

In this study I shall also try to establish the differences existing between these two entities and try to prove that the human being is a creation superior to all other living species.

In the same manner, and notwithstanding the foregoing, in order for all human individuals to be healthy, comfortable and achieve the maximum longevity, they must comply with the laws of Nature designed exclusively for them.

These are laws of morality that govern the internal behavior of human being regarding their nourishment, sexuality and feelings, moderation of excesses.

These laws which we cover in a special chapter, order human individuals to eat a well-balanced diet, to have sex with moderation and to control his passions and negative emotions to attain physical and emotional health.

When a person exceeds the fulfillment of these obligations, damage is caused to the body reducing such person's longevity in addition to damaging the society in which such individual lives.

We shall also cover the ethical laws governing the behavior of the individual in society because ethics are the exteriorization of the moral conduct of humans before the society to which they pertain.

Ethical norms compel the human individual to live in an orderly manner in the society he belongs to, and not to exteriorize their passions negative (anger, violence, vengeance, etc.) and to respect the dignity of their peers to attain social peace.

Now, I ask my readers: what is the origin of passions negative such as hate, ambition,[7] resentment, envy, anger and similar others? Why are they present in the human race?

Passions negative are a part of human nature because they are necessary to defend their survival, but they are resources of a temporary and brief nature, and they are not to be used in daily life due to their destructive power over the individual itself. The constant use of these emotions causes disorder in human societies, as we shall see later on.

In the same manner I will approach the subject regarding the human conscience whose task is to notify the individual about any excess committed against the laws of morality mentioned in the previous paragraphs.

Conscience records all feelings, emotions, and actions of human beings and regulates the functions of internally secreting glands. When the thoughts and deeds of an individual are congruent with laws of moral, his body functions in harmony and equilibrium.

But when the thoughts or actions of an individual do not match such laws, conscience warns him of such infraction in the manner of an "inner-voice" and that person's glands secret excess micro-amounts of biochemical liquids that unbalance metabolic functions and harm the body. Conscience functions are more fully described later on.

---

[7] The excessive search for power, riches or glory.

On the other hand, I want to raise a concern among my readers by asking the following question. Why is the interior temperature of all human beings an average of 98.6° Fahrenheit?

This body temperature is identical in billions of human beings that have populated and still populate the Planet, regardless of whether they live in the North Pole with freezing temperatures below negative 4° Fahrenheit or in a hot desert with temperatures over 104 degrees.

Minimal variation of the average temperature of the body, whether up or down, may be fatal.

Do you think that this is coincidental? In your opinion, what is the reason for this amazing fact? Why is the average temperature 98.6 and not 98.7 or 96.8? Who orders the human body to regulate this temperature?

These questions are developed in the second chapter of this work, and in chapter four I explain the functions of the entity that controls the whole human body, manages heart beat, body temperature, lung movements (breathing), blood circulation, the transformation of food into muscle, bones, blood, etc., and controls also the form and function of each of the internal organs and the outside body shape, which must be identical to the shape of their ascendants.

More than 400 years ago, Descartes affirmed that the last level of wisdom is the knowledge of all the laws of Nature.[8] This reasoning had a powerful influence in my way of thinking, and it was what motivated this work.

Based on that reflection I think that a wise man is that one living in harmony with the laws of Nature.

The laws of Nature are the Will of God regulating the Universe. God is the Author of Legislator of all the Laws of Nature and the Creator of the Universe.

Human beings may believe whatever the want regarding the existence of a Creator; they may proclaim that God does not exist, but that will not affect their lives. But <u>they must know</u> that their bodies and behavior are governed by the eternal, immutable, and mathematically precise universal laws, and that if they do not obey them, they will be destroyed.

All scientists and philosophers that have studied the laws of Nature were right in asserting that those laws regulate the behavior of the laws of physics and chemistry, which are precise, constant, and invariable, and that the human being should must discover and must apply them to obtain their benefits.

However, they have not gone far enough in their studies, I think, because they exclude the laws that regulate nourishment, thoughts, feelings, emotions and behavior of the human beings.

---

[8] Regis Jolivet.- *Tratado de Filosofía. Moral.*- Editor Lohle.- Buenos Aires, Argentina.- Pag. 194.

They do not examine in detail the functioning of the internal organs of all living things obey the eternal and immutable laws of Creation, and this is precisely the goal of this work.

Dear Reader: there is dearth information available regarding the manner in which the laws of Nature affect human beings. If you wish to contribute informed ideas, please send your comments to the following address: p.a.abdo@beinghuman-thebook.com

Notwithstanding the foregoing, in this study I will try to demonstrate that all – absolutely all in the Universe – is subject to the laws of Nature and that by respecting these laws of life we will be better human beings and we will have health, comfort and inner peace.

FIRST PART

The laws of Nature

Chapter I

Historical references

"Nature always maintains her rights, and prevails in the end over every abstract reasoning whatsoever." [9]

- David Hume

For thousands of years, known philosophers and intellectuals have corroborated that the entire Universe is regulated by laws articulable with mathematical precision. These are called the laws of Nature.

The laws of Nature have the peculiarity of being universal, permanent, invariable, and eternal, and they govern the behavior of the energy and the matter through the laws of physics and chemistry.

The oldest antecedent I found with respect to the laws of Nature was from Heraclitus in the sixth century

---

[9] Phrase de David Hume.- Digital Citation: http://akifrases.com/frase/115825

BC. He was the first person we know of to use the word *Logos* in referring to the laws of Nature and used such term in a metaphysic dimension (the study of the ultimate reality).

He held that all things are governed by universal and eternal laws he called *Logos*, the intelligence that directs, orders, organizes and controls the Universe.[10]

In the fourth century BC, stoics continued the thoughts of Heraclitus and sustained that living according to Nature, is living according to the divine order of the Universe under a "common law" because in the Nature lies "the real law shall govern all peoples in all times" because it is "unique, eternal, and immutable."[11] The order of the Universe is due to the existence of these laws.[12]

Stoic ethics affirm that happiness in life is living it according to reason, this is, avoiding passions that are only deviations from our own rational Nature.[13]

Stoic doctrine considers each individual as a part of God and a member of one universal family. It helped to break racial and social barriers and to prepare the

---

[10] Heráclito de Éfeso.- Digital Version.- http://philosophynotebook.obolog.com/heraclito-efeso-205176
[11] El Logos.- Digital Edition: http://www.eleutheria.ufm.edu/Articulos/060316_CAPITULO_II_DE_LA_BUSQUEDA_DE_LA_COMUNIDAD.htm
[12] Ivette Alejos.- *La virtud, el bien y la felicidad en las tradiciones filosóficas socrática/platónica y estoica*.- Digital Edition: http://arje.hotusa.org/griegos8.htm
[13] El Logos.- Digital Edition: http://mb-soft.com/believe/tsxtm/logos.htm

way for the propagation of a universal religion: the Christian.

In effect, before Christian religion, stoics already acknowledged and preached the fraternity of the Human Race and the natural equality of all human beings and in my opinion, this is the origin of democracy.

I return to this Stoic concept and assert, more precisely, that the goal of human life is to live in harmony with the laws of Nature.

To me, the laws of Nature influenced the development of Roman Law, and I consider that they are the basis of present human rights.

Stoics were the ones that further developed the study of the laws of Nature, but they identify them with God, not with His Will, therefore such concept evolved towards pantheist theology of later stoicism.

Logos in Greek and Latin means Word The prologue of the Fourth Gospel mentions Jesus as the incarnation of the *Logos*:

"In the beginning was the Word, and the Word was with God, and the Word was God. The same was in the beginning with God. All things were made by him; and without him was not any thing made that was made. In him was life; and the life was the light of men... And the Word was made flesh, and dwelt among us.[14] John, the evangelist personifies *Logos* in Jesus.

Logos has not been understood by humans even though it has been studied for hundreds of years by renowned philosophers, scientists and theologians

To me, *Logos* is the Will of God expressed in the laws of Nature, and it is the origin of all religions.

Democritus (460-370 BC) developed the atomic theory of the universe. According to his theory, all things are formed by minute pure matter particles, invisible and indestructible, that move through eternity in the vacuum infinite space.[15]

When Democritus imagined the atoms, he considered them eternally stable and representing a universal and unalterable natural law and that an exchange existed between them because they were constantly moving, separating only to make readjustments.

Socrates (470 – 399 BC) linked the laws of natural law with the divine will: "Antigone said to the king who sentenced her to death: I do not believe that your proclamations are more valuable than the non-written

---

[14] The King James Version of the Holy Bible.- Gospel John 1:1-3,4,14.- Digital Edition:
http://d23a3s5l1qjyz.cloudfront.net/wp-content/uploads/2012/09/King-James-Bible-KJV-Bible-PDF.pdf

[15] Biblioteca de Consulta Microsoft® Encarta® 2003. © 1993-2002 Microsoft Corporation. All rights reserved.

immutable laws of the gods, because you are a simple mortal..."

Antigone continued: "they are immutable, not from today or from yesterday, but eternally powerful and nobody knows when they were born."[16]

The laws of Nature, says Marcus Aurelius in the second century AD, "are identical to the Zeus sayings that govern the order of all things. The laws of Nature are the laws of God, and therefore, the laws of reason."[17]

In his second letter to Timothy Saint Paul says, "If we are faithless, He remains faithful; For He cannot deny himself."[18] This means that the Creator adheres faithfully to the laws, and He is not going to vary or contradict them.

In the same manner, Thomas Aquinas (1225-1274 AD), referring to the laws of Nature, said that "not even God himself may alter or disregard these principles..." and he adds that "He is not going to

---

[16] Sócrates.- *Diálogos de Platón*.- Digital Edition: http://vagabundeoresplandeciente.wordpress.com/2008/04/10/antigona-de-sofocles-y-la-distincion-juridica-cardinal/
[17] José Antonio Ñique de la Peña.- *El Humanismo Jurídico en San Marcos*.- Digital Edition: http://sisbib.unmsm.edu.pe/bibvirtualdata/tesis/human/%C3%91ique_pj/pdf/parte1.pdf
[18] The Bible, supra, at note 14.- 2 Timothy 2.13.

change the order of things... not even when it is a matter of justice between a man and another man."[19]

Averroes (1126-1198), Hispanic-Arabian physicist, theologian and philosopher says that: "God created Nature, He gave it a physical order and He established mathematical laws."

Concluding that "after that He left him [mankind] to act freely, abstaining from interfering with revelations or miracles and He did not interfere further."[20]

Notwithstanding the foregoing, there have been amazing healings called "miracles" which laws – at least up until now – have not been deciphered by human knowledge, but based upon the immutability of the laws regulating all Creation, I can assume those laws shall be discovered eventually.

Larry King (an American reporter) asked Stephen Hawking, the British philosopher with a doctorate in mathematics who is also an atheist (as commonly defined), the following "Do you believe in God?" to which Hawking responded "Yes, I do, if by God you mean the embodiment of the laws that govern the universe."[21]

---

[19] Santo Tomás.- *La Suma Teológica*.- 1-2, 100, 8, ad2.- Digital Version: http://es.scribd.com/doc/18678110/resumen-santo-tomas-de-aquino-suma-teologica
[20] Marcelino Cereijido.- *Ciencia Sin Seso*.- Siglo xxi Editores, S.A. de C.V.- sexth edicion 2005.- Pag. 204.
[21] Interview by Larry King of Stephen Hawking, December 25, 1999 at 9:00 pm, on CNN television: http://www.psyclops.com/hawking/resources/cnn.html

The perception that humans have over the laws and the order of Nature, as well as the immutability, permanence and precision of its laws, has been proven by Science. We will study this in the chapter dedicated to Science.

I am including these historic references so the readers has more arguments to help realize that although these laws have been known for centuries, human beings have ignored them by not applying them in their ways of thinking and in their actions towards others.

In the same manner I do so to invite the reader to accompany me in this investigation analyzing the transcendence that the laws of Nature have in the conduct and behavior of humans, as by understanding the same, a reader may be able to understand this study. More significant, however, is that by applying these laws, we may significantly improve our health, emotional balance and relationships with others, and conserve the environment.

Chapter II

The laws that govern Nature

"True law is right reason in agreement with Nature; it is of universal application, unchanging and everlasting... [and] will be valid for all nations and all times."[22]

- Cicero

Since appearing on Earth, all we humans have realized that within the environment surrounding us and within the movement of the celestial bodies (the Universe) there is an order that repeats in an unchangeable manner in all actions of Nature.

Anywhere we find the existence of order we can conclude that laws exist and that said order cannot

---

[22] Marco Tulio Cicerón.- *De Republica*.- Digital Version: http://www.eleutheria.ufm.edu/Articulos/060316_CAPITULO_II_DE_LA_BUSQUEDA_DE_LA_COMUNIDAD.htm

be the result of chance but the direction of the one Legislator who established it.

In this work's prelude, I commented that the behavior of the so-called natural phenomena are not mere chance, but the result of a systematic respect of certain laws. Those laws are called the laws of Nature.

The laws of gravity, thermodynamics, reproduction, life, inheritance, physics and chemistry, electricity and magnetism, etc. are all laws of Nature.

In the same manner, natural phenomena such as wind, rain, tides, etc., do not act per se. They are governed by those laws that operate in coordination with one another.

The laws of Nature are a group of unchangeable, permanent, universal laws that govern the behavior of energy, matter and all living things, as well as all natural events and processes. The laws of Nature govern the Universe.

Any deed, process or Natural phenomenon has a previous cause. Neither luck nor misfortune exist in Nature because the Universe is planned and ordered by these laws. Nothing happens by chance, there is a master plan that can be deduced from the order that results from such laws.

The Stoic Lucio Anneo Seneca (4 BC to 65 AD) affirmed that all events if the world are rigorously determined therefore "freedom lies in the acceptance of our own destiny, which consists, fundamentally in living according to Nature."[23]

The stoics compared the laws of Nature to a dog tethered to a carriage. When the carriage advances, the dog has no other choice but to follow behind but it can do it in two different ways, by accepting and following it or being dragged by the neck suffering all the way.

This kind of reasoning leads us to understand that despite our philosophical, religious, metaphysical or atheistic beliefs, we have the obligation to obey the laws of Nature and to accept with valor and resignation our emotional or health problems when they have no solution.

Stoicism affirms that life is the result of our actions and that fate does not exist, but it is the result of ignoring the causes of events. If our minds are able to accept all connections of the causes, we may understand the present and predict the future.[24]

Events and natural phenomena such as earthquakes, hurricanes, draughts, floods, plagues, etc. are neither good nor bad. These classifications are given by humans because man considers it so as adversely affecting or benefitting humanity in some way.

---

[23] Séneca.- *De la Felicidad.*- Digital Version:
http://www.buenastareas.com/ensayos/De-La-Felicidad-S%C3%A9neca/1252522.html
[24] Estoicismo.- *Wikipedia.*- Digital Version:
http://es.wikipedia.org/wiki/Estoicismo

But what is Nature to you, dear reader?

To me, Nature is everything that spontaneously exists in Creation that human beings have not intervened with in order to exist.

Nature is everything we are able to identify as Universe, and we understand as Universe all energy, matter, all living things, all deeds, processes and phenomena occurring therein.

Cosmos is a broader reference for Creation. Universe and Nature are more specific in respect to what they cover within human terms.

Thus, I can say that Nature is the specific form of what we humans determine as pertaining to Creation.

We can speculate that the laws of Nature were dictated before the Creation as specifications.

To understand the foregoing we can imagine that Creation was made in three stages, during the first one, the Creator designed the Universe; during the second stage He established the laws for such Universe to work according to the design, and during the third one He created the Universe according to the specifications including, of course, the living creatures.

This compels me to think that no laws may exist without someone creating them. I deduce from this, that an omniscient, omnipotent, Legislator of infinite Nature dictated such laws.

It also makes me conclude that everything surrounding us is Creation, not fate, and necessarily the creation of someone, even though it cannot be all explained by us because something cannot come from nothing.

To me, atheism has never been reasoned correctly, it has only been limitedly professed to deny the existence of a Supreme Being.

The most prodigious work in Creation and the most important law of Nature is that of time; the Universe would not exist with it… nor you and I, dear reader.

The Universe respects time. Time works in coordination with movement. The Creator designed these energies so that they function in an orderly way. Order is the object of time.

Base don reason I can affirm that time was created before the so-called Big Bang, that is to say, before the existence of the smallest part of matter.

Time is the same, identical, in all the processes of Nature: growth, development of plants, animals, microbes, human beings, winds, rains, earthquakes, etc., without regard to what part of the world is affected besides time works the same way in inert matter as in organisms.

To comprehend this, I consider it necessary to open a parenthesis and first define what I understand by "intelligence"; it is basic for all this book attempts to deal with.

The need to define words and concepts is indispensable if we want to express our ideas with more clarity and to avoid, if possible, ambiguities.

What is the purpose of intelligence in human beings?

This question has been formulated to many people at various cultural levels. The majorities of them become confused, stammer and have no proper answer at hand.

To help them, I ask again, what is the purpose of intelligence in a tiger or a lion in the jungle? Most of the people surveyed answered that animals have no intelligence but instinct only. Then they give me an answer: the purpose is to procure food and seek survival.

In effect, the main order of Nature is to live, and intelligence is an instinct that serves an individual of any species to achieve said purpose, even though its degree depends on the zoological scale.

In that sense, I dare to give this definition not found in the dictionaries or encyclopedias: Intelligence is an instinct that all living things have that gives them the capacity to process information received from their surrounding environment by means of their sensorial organs to procure survival and comfort.

I must clarify that the intelligence instinct –I call it genetic intelligence– that animals have, is an outline or trace of the intelligence developed by human beings because it serves them only in seeking survival and comfort, not in solving problems or understanding ideas.

In addition to the foregoing, we have no elements, either to collect and evaluate the reasoning of animals or to identify whether their mental qualities are intuition or reasoning.

Said animal intelligence-instinct is present also in plant and unicellular organisms that have overtones (or rudiments) of intelligence in tropisms or psychisms to seek food. Since intelligence is an instinct it is universal for all living creatures and therefore genetic.

But human beings have a superior level of intelligence to other living creatures. In addition to the capacity given by the intelligence-instinct mentioned in the previous paragraphs, human intelligence can be used to reason correctly, seek longevity, look for and analyze solutions, solve problems, discover and apply the laws of Nature for human benefit and for the benefit of other living species.

I would like to explain the term longevity as the maximum period of life for which a living creature was created. Longevity is concomitant to survival, this is, they act jointly.

It is important to know that the search for longevity is exclusive to the human race. The reason for this affirmation is that the length of life in all living creatures has been incorporated into their genetic codes since it programs all their activities and it may not be prolonged. This it what I call genetic longevity.

But the human being has the power to prolong his genetic longevity to try to live the most time possible, and to that end, humans try to balance their food intake, manufacture medicines to cure illnesses, take food supplements, exercise, and establish compulsory safety measures for automobile drivers and house constructors, etc., as we shall see in the chapter dedicated to the spirit.

Additionally, intelligence in human beings is located in the brain, and I can infer that superior animals and many other inferior ones, have a brain in which intelligence is located too.

We can find nervous systems performing brain functions in insects, microbes and other animas lacking a "brain" as commonly defined, and there, I consider, is where we can locate a pseudo-intelligence.

On the other hand, I wish to tell my readers that Nature has its own unchangeable, permanent and universal rules by which is governed.

From this arises the following question I would like to formulate for my reader: which process, by chance, is able to explain the complex optic system where drivers transmit simultaneous signals from the eyes to the brain, where luminous impulses transform into images captured by the conscious mind at the same time it produces ideas?

I can answer, unequivocally, that the optic system is originated when a sperm of nearly infinitesimal size fertilizes an egg and in that moment, not only the optic system is originated, but all the other apparatus and systems in the human body –digestive, skeletal, circulatory, muscular, etc.– and have remained unchanged in the millions of human beings that populate and have populated the Earth.

These systems and organs develop through the years in an identical and invariable manner in the body of all human beings, which body is designed to have 32 teeth, 206 bones, 2 kidneys 20 fingers, etc.

The fact that some organisms do not have fewer or additional pieces does not mean that their design has been altered. Eventually the casuses of such things will be discovered because the laws of Nature are invariable. Moreover, these differences are not passed on by inheritance, but the descendants of those with less or more members will still have the normal design of all human beings.

The laws of Nature were created by the Intelligent Matter, by the Creating Evolution, by Nature or by God or whichever name you wish to call the Creator. This is just a matter of terminology. The fact is that the Universe and all living creatures existing herein have been designed for the previously created environmental conditions.

The human being can profit from laws of Nature even if they ignore them; if human beings take advantage of said laws they shall benefit from them but if they do not take advantage of such laws no benefits will be received, and if they are not used properly or they do not comply with their terms, whether consciously or unconsciously, they shall suffer because any omission of a law or excess in its fulfilment results in a penalty, so to speak.

Just as breaching any law carries a penalty. Jurists consider that a law without a penalty is like a bell without a clapper. If the violation of a law carries no penalty, it cannot be called a law, it would be, at best, a rule, a letter, a manifest or a decree, in short, a suggestion.

All laws governing Nature are of a mathematical precision and as an example we have the ones governing physics, chemistry, electricity, biology, and a long list of areas. We consider that there are thousands, if not millions, of laws that govern the Universe and we are unaware of most of them.

Now, in order to comply with the first commandment of the laws of Nature, survival, all living things are compelled to obey the specific laws that rule their own species and that are incorporated from their conception into each one of their cells.

The laws of Nature are not written like the civil laws dictated by human society, but these are incorporated into the genetic code of each species and the human being has the obligation to discover them through observation and reasoning.

We may compare the genetic code of one organism to something like a computer. Each species has its own software, so to speak, that covers its vital functions and obligates it to behave in the same manner as in all other bodies of the same species.

In other words, the genetic code of all living things is recorded in a software containing all instructions that should be followed by the body while living and it is transmitted to its descendant in an identical and invariable manner because it is a law of Nature.

Now, with respect to human beings, each cell contains all information governing its individual behavior and the conduct of the body to which it belongs, this is, all cells that form an organic apparatus (digestive, circulatory, respiratory, etc.) perform their own duties and the ones of the whole organic system of which they are a part, (muscular, nervous, skeletal, etc.), and all organic systems, in turn work individually and jointly with the being to which they are a part, functioning in perfect harmony.

In obeying these invariable laws of Nature that apply to human bodies, we can conclude that said laws affect in the same manner all individuals whether young or old, rich or poor, healthy or sick, wise or ignorant, Kings, Presidents, Popes, etc., since to the laws of Nature there are no superior human beings; they are all the same.

I also consider that the births of children with congenital defects are not "errors" of Nature but due to actions of the parents, as from the use of contraceptives, drugs, improperly-treated gonorrhea, accidents during pregnancy, and dietary deficiencies, among others, because all that happens in Nature has a cause and a reason.

Even while there may be no conclusive evidence, these illnesses may be caused by, among other reasons, conceiving in an improper age as I consider that the laws regulating reproduction of the human species are designed for when the body has maturity to conceive, neither before nor later.

My main premise is that the laws of Nature are invariable, permanent and eternal and that this has been demonstrated thousands of millions of times with children that have been born during centuries and still are being born perfectly healthy, therefore I can conclude that eventually it will be discovered that those deficiencies were caused by humans, not by Nature.

The laws of Nature order that an egg be fertilized by a sperm of the human species always. Infallibly the product of such fertilization must be a human being, and it is as of this moment when the new molecule has all the information that it requires it to function.

I also consider that upon conception the manufacturing of the new physical structure (the body) is initiated, and it begins to take the shape that each internal organ will have (heart, kidneys, lungs, etc.) as well as the outside shape of the human body, obeying invariably the program incorporated into the genetic code of each one of its cells.

Also, from the moment of fertilization, all internal functions of the newly developing organism begin, such as digestion, nutrition, metabolism and blood circulation, including the constant regeneration capacity and the expulsion of thousands and thousands of waste cells that are of no use whatsoever to that being.

Now, from the foregoing I wish to ask my readers the following question: what parameter do we need to use to classify a fact as a law of Nature?

I think that **everything that is universal in all human beings must be the parameter to be used to consider a fact as a law of Nature.**

Menstruation, for example, is the programming in all the women of the world for the reproduction of the species and since it is universal, it is therefore a law of Nature.

Actually, some scientists try to eliminate women's menstruation. They wish to modify the human body so that women stop having the inconvenience of this process, but they do not realize that menstruation is the basis of life, therefore modifying this process may have unpredictable negative consequences in the women's body.

Deducing from the previous, I can conclude that the modification, interruption or elimination of any process of Nature is going against the Creation and since Creation is life, is going against life.

Civil laws should be enacted to favor compliance with the laws of Nature. The governments of the world should know (whether they understand where said instruction is coming from or not) that they must procure that all constituents, without exception, live for the longest time possible whether they are homeless, drug addicts or felons, sick, disabled or terminally ill, because it is so ordered by the laws of Nature. The greatest longevity of the people of a country should be the goal of their government, and all of the governments of the world respectively.

Notwithstanding the foregoing, some people consider they have the right to dispose of their lives or health freely, because as they try to explain, they are not obligated to prolong their longevity and they are not harming or invading the rights of third parties and they are the owners of their own life and therefore they can do with it whatever pleases them.

Some drug addicts and other suicides claim they have the right to their own lives, in the same manner as civil laws authorize men and women to dispose freely of their own possessions, somehow invoking the rights given by society.

In my opinion, this way of reasoning is erroneous because it ignores that the laws of Nature do not grant any rights, let alone property rights over individual lives, but rather the laws of Nature compel living driven by the survival instinct and there is no freedom to break such laws.

All human beings are programmed to live one hundred and twenty years of age.[25] That is the age that scientists have proven that may be achieved by all humans, besides, the fact that some people have achieved such longevity means that all human beings are capable of reaching it because we have been created equal before the laws of Nature.

Genesis[26] mentions that God declared: "My Spirit will not always strive with men, for that he also is flesh: yet his days shall be an hundred and twenty years."

Searching for human longevity is the reason governments impose compulsory safety measures on housing or building constructors and on the manufacturers of machinery, equipment and vehicles, thus trying to protect the lives of their constituents.

---

[25] Ciencia y Salud.- Longevidad humana, ¿existe un límite? Cita Digital: http://cienciaysalud.laverdad.es/4_4_21.html Límites de la vida humana: longevos y grupos de longevos.- Digital Version: http://www.seme.org/area_pro/textos_articulo.php?id=9

[26] The Bible, supra, at note 14.- Genesis: 6:3

They also limit the freedom of their citizens by compelling them to use safety belts in their cars and to respect the speed limits.

They try to eradicate crime, drug addition, illnesses, and achieve peace among society. They do all this so inhabitants of a county have a longer average (longevity).

Progress and longevity in human society is due to the discovery of the laws of Nature by science and their application by means of technology as we will see in the chapter dedicated to Science.

Now, I ask my reader, what is the parameter we need to use to determine the progress or regress of a country?

According to what we mentioned in the previous paragraphs, the life average (or life expectancy) of all people should be higher every day.

In my opinion, we must use this parameter to determine the progress or regress of a country in the fulfillment of the laws of Nature.

Presently, the life expectancy in developed countries is high, while in underdeveloped countries it is low.

But, taking into consideration that the laws of Nature are universal and apply to all human beings without distinction of race or nationality, all humans should help each other in order to achieve the longevity for which they were created.

The obligation of the most powerful countries should be to support the weakest countries so they achieve the same level of development and the same longevity because this obligation is commanded by the laws of Nature.

History has proven that if rich countries fail to comply with this obligation, eventually they will fall to the less powerful ones because the less powerful sense that their impoverishment is caused in some manner by the abuse of the rich ones.

So poor countries keep fighting constantly to be heard and vindicated in their rights granted by the laws of Nature for their nourishment, work, housing, and recreation.

Chapter III

Science

Real Science is based on experience, reason, the laws of Nature, mathematics and experimentation.[27]

- Leonardo Da Vinci.

In order to use the laws of Nature to the benefit of the human race, people must go to systematic, statistic reviews of Natural events, and that requires Science.

All facts and natural phenomena obey its rules and the sole purpose of Science is to study and prove them by a systemic, theoretical or practical method or process.

Spanish language dictionaries confuse "Science" with "knowledge." Everything known to the human being is knowledge, but not all knowledge is Science.

---

[27] Escritos Sobra la Ciencia.- Leonardo Da Vinci.- Digital Citation: http://es.slideshare.net/rurenagarcia/escritos-de-leonardo-sobre-ciencia

Therefore to me, **Science is the method or process used to investigate and prove a law of Nature. Once the law is proven, it is enunciated and then it becomes a part of human understanding (knowledge).**

To demonstrate the validity of a scientific principle or a theory, scientists have to experiment once and again passing through rigorous trials of the physical, chemical or biological processes that are under review, keeping constant all factors intervening in said processes and obtaining always, infallibly the same positive results, and then, only then, it may be given the name law of Nature.

A law of Nature is valid for the human knowledge, when it is true; this is, when the relationships referred to, or expressed thereby, happens. Therefore it is indispensable that all facts confirm such law through practical or theoretical experimentation by mathematical formulas.

The proof must be complete and absolute. One simple failure, that is, one exception in the proof of the law definitely voids that law in that formulation as a scientific principle.

The human being must investigate, discover, obey and apply these laws to obtain benefits in order to preserve survival and achieve longevity.

Additionally, I can deduct that the laws of Nature can be discovered, both in their experimental and theoretical manner.

Experimental scientists may, through observation of Natural facts or phenomena, discover the laws governing them and express them in mathematical formulas or through practical experimentation by operational methods.

Theoretical scientist, in turn, may by reasoning, deduction, induction, and mathematics, discover the laws of Nature, but they have to empirically prove and corroboration their investigations in order to be converted into Science.

Newton, as one example, was an experimental scientist. He discovered the external laws of matter by observation of Natural phenomena. Within the laws that he discovered are universal gravitation, body acceleration, thrust, light decomposition, body attraction, etc.

Einstein, in the other hand, was a theoretical scientist who discovered the internal laws of matter by reasoning, deduction and mathematics. Within the laws he discovered are the behavior of atoms, energy, matter, etc.

But… What is truth for you, dear reader?

Truth is the coincidence between what we think or believe and what is real. Science is the demonstration of such reality.

There are three ways to reach knowledge and truth.

- Truth is objective when obtained directly by observation and it does not require proof to be understood, for example, movement, day and night, time, etc.

- Truth is historic when it is proven with documents, archeological data, pictures, testimony, engravings, books, newspapers, etc.

- Lastly, truth is scientific when proven infallibly by practical or theoretical experimentation through mathematical formulas.

I think that there is only one truth. It may not be two truths as it happens with the feeling of certainty or beliefs that may differ for each person and may or may not coincide with reality.

For example, in the old days people had the certainty that the Earth was flat or that Earth was the center of the Solar System, nevertheless, said beliefs were untrue because they did not match reality even though the idea would have been of universal application.

Scientific theories are not Science, they are only that, theories, but if they are proven experimentally or by mathematical formulas, with constant positive results, they turn into a scientific fact. Scientific facts are what we call Science.

There is no democracy in Science. The opinion of the people does not count; scientific facts are the only truth. In the year 1931 a book was published in Germany called "One Hundred Authors Against Einstein" compiling their opinions contradicting Einstein's opinions with the purpose of discrediting his investigations because he was Jewish. When he was asked about this he responded: "Why one hundred authors? If I were wrong then one would have been enough!"[28]

The laws of Nature are the only and absolute truth that may be demonstrated by men and Science is the only infallible process that men have to find truth.

Some people have asked me why I do not include Revelation in this book. I must say that this work does not include religious subjects because they cannot be proven by experimentation as the laws of Nature, which can be demonstrated by Science once and again in an infallible manner always obtaining positive results.

Human's beliefs are not laws of Nature because they are not universal, since they vary for each human being and each one has his particular way of

---

[28] Las Anécdotas de un Genio.- Digital citatio:
http://debates.motos.coches.net/archive/index.php?t-244734.html
http://es.wikipedia.org/wiki/Cien_autores_en_contra_de_Einstein

believing or thinking and they cannot be demonstrated empirically.

There are no "exact sciences." Only the laws of Nature discovered by Science are exact. Scientists support their investigations with mathematics, statistics, deduction, induction, logic and reasoning to prove their discovery of a law of Nature.

Scientific determinism is a universal principle (also called the causality principle or sufficient reason principle) affirming that all events and all happenings are caused; all happens for a reason, sufficient cause or by necessity; nothing happens at random.

Albert Einstein used to say "God does not play dice. I believe in complete law and order in a world that objectively exists"[29] indicating that that nothing happens randomly in the Universe because Nature has an order since it is subject to the laws of mathematic precision and that even what we humans perceive as disorder is not because it keeps the logical pattern that will be demonstrated eventually.

Through logical and mathematical reasoning we will be able to discover a principle or a law, which is enough to establish a theory, but this shall not become an object of Science, per se, until it is proven again and again to be absolute and faultless through experiment

---

[29] Albert Eintein.- Citas.- Digital Version: http://es.wikiquote.org/wiki/Albert_Einstein

or mathematical formulas. **Science is the truth demonstrated through experimentation.**

Logic is the substantiation of the laws of Nature. Mathematics is the language, the manner to express the laws of Nature to attain its control, application and it use by the technology.

Mathematics does not exist. As a language, it comes from the human imagination and helps us to express concepts for better communication. In other words, we cannot understand the meaning of the number four, considered only in the abstract. But if we say four meters of four pounds or four dollars, etc., then we can understand the number and we get the idea, and are then able to reason over said issue.

During the Descartian era, all knowledge was called Science because philosophers were at the same time scientists, astronomers, medical doctors, poets, etc., but as of the eighteenth century, Science started to be defined and to displace philosophy and scientists began to investigate, discover and demonstrate the laws of Nature expressing principles and technologists began to apply said principles producing foods and medicines in a large scale, as well as machinery, fuel, equipment, clothes, etc.

Science investigates, discovers and determines the means to control a law of Nature and technology applies the laws discovered by Science to manufacture consumer goods and amenities for the survival and longevity of the human race.

Science, through technology, has improved the quality of life for human beings. It has made large-scale production of food possible to nourish Humanity.

Science has liberated human beings from illnesses and from hard work and has provided different devices for our comfort and entertainment, aiding us to achieve the longevity for which humans were created.

The production of beer, glue, plastic, medicines, means of transportation, fuels, and clothing among others, are scientific processes developed by technology.

When primitive people started rubbing flint stones over wood char to produce fire, they were developing scientific fact. Every time they did that process obtaining the same results in the same conditions, they were demonstrating a scientific fact. This primitive man or woman was a scientist. No title is needed to become a scientist.

The scientific fact and scientific process, happen in every field of knowledge: Chemistry, Physics, Biology and in general in every Natural phenomenon.

For example, if Science occurs in Chemistry, we should not call it Chemical Science because Chemistry is a field of study, not science itself. <u>Science is the process or method</u> used to discover such laws.

We can say the same if it occurs in Physics or Biology. Nevertheless, the laws regulating Chemistry, Physics or Biology can indeed be called laws of Chemistry or Physics or Biology.

I shall define Physics as the laws of Nature governing phenomena spontaneously occurring between matter and energy in space.

In the same manner I shall define Chemistry as the laws of Nature governing the transformation spontaneously occurring in the inner structure of matter on the molecular order.

And I shall define Biology as the laws of Nature studying living creatures as of their molecular structure.

I think that the ill-named juridical, social, economic, sciences which principles cannot be demonstrated empirically should not be called Science. Science is a method or a process.

There are no different "sciences" as asserted by some, but rather there are scientific disciplines using proprietary scientific investigation methods such as deduction, induction, statistics and logic, but they should not receive the name of Science because their study or principles have not been substantiated with positive results in an infallible manner.

No definition of Science is to be made ignoring the laws of Nature; therefore I dare to assert that

Science is the only true way that humans have to know the will of God expressed in the laws of Nature and to understand the eternal plan created for the world and its residents.

Now we find ourselves with another question: Can we be sure, without a doubt, that God exists?

To answer that question we must think of the nature of energy itself.

If we give the energy or the infinite force that created the Universe the name God, Evolutionary Creation, Chaos, Nature, Fate or any other name we wish, we can still prove His existence even if it be incomprehensible by the deepest human thoughts.

Then, the Universe, and particularly the events occurring on Earth affecting living things, matter and energy, are all subject to the invariable, permanent, universal and eternal laws that create the order of things.

Taking into account that order cannot be the result of chaos but of law, we conclude from the foregoing that there was an Omnipotent, Omniscient Legislator or Author of an infinite nature, who made the laws of Nature.

In the same manner, humans in general, not only scientists, are able to observe our surroundings, our ecological environment and be aware of the Creation we are a part of, and that this Creation also obeys the laws that originated in the same entity of infinite proportion whom we call God. The human mind is finite and not capable of understanding the infinite.

God is inaccessible, inscrutable for humans, but we can know him through His manifestations; the laws of Nature and the works of Creation.

Science can discover what is true and what is good, fair or wrong regarding the human being.

Humans have thousands and thousands of reasons for their development and progress. It is necessary to discover them to later apply them. The laws of Nature govern these reasons which are part of the Universal Reason, which is Creation.

The laws of Nature have Creation as their universal reason.[30] Reason is the ordering principle of the Universe. Human beings have the gift of the capacity to reason.

Now, I ask my readers again, why do we say that God is good if God is an unintelligible entity?

I consider that a reasonable answer would be because all laws of Nature were created to benefit the human race; and for me, that is good, that is love.

---

30 Reason: cause, reason. It's name. Reasoning: it is a verb. There is no relationship grammatical among them.

As far as Science has been able to prove, the only purpose of existence of the human being in the Earth is to obey the laws of Nature that are the will of God.

In other words, the will of God is the reason the laws of Nature exist, therefore, to obey His commands we must comply with such laws and incidentally we will have a life full of health, harmony and comfort.

Chapter IV

The laws affecting all living creatures

The Soul and the Life

Things with souls differ *from* those without souls *in being* alive.[31]

- Aristotle.

The soul is one of the concepts least studied by Science because it is abstract and subjective and as intangible as intelligence or reason, it is hard to specify or define, because it cannot be measured objectively, but, by the specific functions it performs we may define and determine its practical scope.

---

[31] Frases Célebres de Aristóteles.- Digital Version:
http://www.erroreshistoricos.com/frases-historicas/380-frases-celebres-de-aristoteles.html

An old saying proclaims that life, breathing and body movements come from the soul.[32]

Aristotle (IV Century BC) thought that the soul is the substance that makes the body a living thing and is inseparable from the body.[33]

Greeks considered it as "a principle of internal life residing in all living organisms allowing and regulating both the physiological and mental functions."[34]

To this date, the concept of soul is used frequently in the religious sense to express something divine, immortal and exclusive of the human being.

This idea of the soul is shared by many people and each one's ideas might be confused with the ideas included in this book because I adopt part of its original meaning because I think it has some truth to it.

Due to the foregoing, I will try to give objectivity to the original concept of soul excluding it from the religious concept, based upon the functions attributed to it by consensus.

First, I must state that all living creatures possess qualities that distinguish them from inert matter. These qualities are the capacity to nourish themselves, grow

---

[32] Planteamiento Antiguo: Alma y Cuerpo.- Digital Version:
http://platea.pntic.mec.es/~macruz/mente/cuerpo-mente1.htm
[33] El alma en Aristóteles.- Digital Citation:
http://filosofia.laguia2000.com/filosofia-griega/el-alma-en-aristoteles
[34] Planteamiento Antiguo: Alma y Cuerpo.- Digital Version:
http://platea.pntic.mec.es/~macruz/mente/cuerpo-mente1.htm

and reproduce because they have their own movement and they are continuously renovating themselves.

From where do you think, dear reader, come those instructions ordering the living creatures to eat, grow and reproduce?

Who informs matter (the body of a living being) to constantly perform these activities in a scheduled manner without rest?

These activities do not act by themselves in a living being nor are they driven from the outside because their body is not a machine; necessarily an entity must be inside said creature –other than the body, constantly managing it, and not letting it rest.

To me, the information controlling a body in order to function comes from an immaterial and subjective entity called the soul.

All living things are composed of two elements: the body and the soul; the latter is considered as the "organizer" of the matter constituting an organism.

A body without a soul is not organized; it is inert matter. Only living creatures are organisms, which is why it is redundant to say "living organisms."

I can define the soul as the energy that provides an organism with all the information it requires in order to function.

In other words, the set of instructions (or program) incorporated into the genetic code of each cell of a living being regulating its shape (both internal and external) and the functioning of its organism during all its existence, comes from the soul.

In my opinion, this is the only explanation I find rational regarding the origin of these activities (or capabilities) of living beings because the existence of this information is generic and it is demonstrated by the objective and reasoned analysis provided by experience.

The fundamental function of the soul is the survival of the individual, because the soul manages all the laws producing the basic, or physiological, needs vital to an organism such as eating, drinking, reproducing, sleeping, etc., as well as the sense organs of sight, hearing, touch, etc.

These laws of Nature are called "instincts" and they are programmed in the genetic code of each individual, as we shall see in the chapter dedicated to the instincts.

The soul is complemented with the life of the living things. **Life is the energy that activates or makes the genetic program (the sole) of each cell of an organism function.**

How does life originate? We do not know, but we do know that said vital energy, as well as the soul, cannot be reproduced by men, but is passed by inheritance.

The soul and the life are two laws of Nature because they are universal and they regulate all living creatures.

These laws, inasmuch as they regulate conduct and behavior of living beings, we may call "laws of life."

In the beginning, the soul is leading the manufacturing of the physical structure (the body) of each species, it determines the shape of each one of the internal and external organs every living thing will have; the internal and external shape must be identical to the ones of their ascendants.

Afterwards, the soul leads the invariable functioning of each one of their internal organs while the creature is alive, constantly renovating them because the organism does not rest and keeps the shape of each one of them and it also heals them when they are injured.

Now, in regards to the human being and from the functional point of view, its physical structure (its body) hosts two nervous systems with different and independent functions, but so intimately related that one could substitute for the other.

The first one, the autonomic nervous system, is composed by the sympathetic and parasympathetic system that is in charge of the unconscious processes of your inner world. This system is not controlled by the will of the human being; thus it is called autonomic.

The sympathetic nervous system is in charge of preparing the body to react to a stressful situation, but, the parasympathetic nervous system keeps the body in regular conditions after the stress situation has passed; this is antagonistic to the sympathetic one.

As an example I can say that when an individual listens to the shot of a firearm, the sympathetic nervous system acts immediately preparing the organism for action and to defend its survival by increasing its force, the heart beat frequency, contracting its muscles, increasing its breathing, dilating its pupils, etc.

But once the stress situation has passed, the parasympathetic nervous system starts functioning to inform the individual that danger has passed and reinstates the organism to its relaxed normal condition.

The latter of the above mentioned nervous systems is the central one, and its function is to put an individual in contact with the outside world through sensory organs (sight, hearing, tact), and it is controlled by the will.

The soul acts only in the internal organism of human beings through the autonomous nervous system, composed of the sympathetic and parasympathetic nervous system.

This means that the soul, through this systems, has the full control of the organism of a living being directing all of its internal functions like nutrition, digestion, respiration, metabolism, blood circulation, development and reproduction. With this I want to say that the soul is in charge of all the internal processes of the organism.

All this functions also are managed by the soul through its control over the internal organs and apparatus (heart, lungs, kidneys, etc.) and without the intervention of will, wishes or intelligence of the individual, this is, the organism keeps functioning even when the individual is asleep, unconscious or in a coma.

Some of the laws of Nature that are under the control of the soul and that regulate human body functions are:

- The laws that regulate the basic or vital physiological needs to achieve its survival, those are, fundamental functions, such as eating, drinking, sleeping, breathing, excreting, etc.;

- The laws that regulate the functioning of the sensory organs that allow the individual to make contact with the outside world, such as sight, hearing, touch, etc.;

- The laws that regulate the unconscious vital processes of the organism, such as blood circulation, digestion, metabolism, the body temperature, etc.;

- The laws that regulate the inner and outer body shape;

- The laws that regulate the reproduction of the species.

Now, I can mention as an example of the what I said before, that at the moment a human being of any age, race, gender, or social status, puts food in his or her mouth, all internal laws of alimentation that govern the survival of the individual begin their functions

Therefore, a part of the food consumed by the individual shall be converted by his body into blood, some other in muscle, another in bone, skin, nails, hair, etc.; it regenerates cells and rejects thousands and thousands of them that are not productive anymore and will discard through urination and defecation whatever is unusable.

The human being must not interrupt the feeding process because any interruption in this process of Nature will harm his own body.

As an example, I can mention Bulimia, when a human being looks to eliminate the excessive intake of food through vomit or laxatives.

In the same manner we can speak of the laws that regulate the reproduction of the species. Human beings must not interrupt the reproduction process either because from the moment the spermatozoid enters the woman's body the laws of creation that regulate reproduction begin to work, and that minuscule spermatozoid goes in search for an egg –knowing its location because it follows the instructions of its genetic code- fertilizes it, and then the gestation process initiates independently of the women's will or desires a process which is identical in billions of humans that populate or have populated Earth.

Here I repeat what I have mentioned previously: Any interruption to a Natural process, such as an abortion, goes against life, including the life of the mother, because the interruption of a law undoubtedly will have a sanction, even if not present at the time of the transgression.

Soul and life are two different energies in function but they depend on each other. Soul and life are concomitant and complementary in the animation and internal functioning of living beings.

Science and technology have created computing systems that pretend to copy or imitate the human model.

Making an elemental analogy of the human body with computers, I can make a scheme of it as follows: A computing system consists mainly in a mechanical structure (hardware) that will function or better said, will give the answers to the information given through a program (software) that gets the whole system to work and that we consider an "operating system".

The analogy with the human being, briefly, is, that the "hardware" is the human body (the structure), and the "operating system" is the soul (the software or program).

Up to here, neither the computing system, nor the human model work. The information of the operating system to be processed by the hardware is there, but these two elements do not get to work if we do not energize them. The electrical energy that we connect to the computing system is the equivalent to the life that the human body receives.

As a conclusion to the forgoing I can say that soul without life will not be able to inform the human body to perform all its programed functions. Life and soul are the two energies that the human body requires to live.

On the other hand, soul only acts over the organism's matter when it has life. All organisms are animated by life and soul. Life and soul disappear together from our reach when the individual dies; they are not immortal.

I make this last affirmation because the concepts of soul and life have been known for thousands of years and are considered to end with the death of the individual.

Indeed, Greeks, as in many other primitive cultures, understood the soul fundamentally as the principle of life of every human being; that such breath or vital principle is in all living beings is not eternal, because it disappears when the body dies.[35]

These concepts are congruent with the definitions of soul and life which I argue in this work, since I consider that these energies are only active when organisms have life because its function is, precisely, to give life to them, and this is the reason I affirm that they disappear when the individual dies.

Irrespective of the foregoing, you, dear reader, can give the name you wish to these entities that we commonly call soul and life, because the important thing is to have a word that identifies the entity that controls the inner and outer form of all living creatures, as well as the invariable functioning of their bodies while they live.

---

[35] Filosofía Griega.- Origen de la Filosofía-Presocráticos-Sofistas y Sócrates.- Digital Version:
http://www.e-torredebabel.com/Historia-de-la-filosofia/Filosofiagriega/Presocraticos/Alma.htm

SECOND PART

Laws governing the human being in particular

Chapter I

The Spirit

"The spirit shall speak for my racee"[36]

- José Vasconcelos.

People, with a conscience and use of reason have asked themselves, Why do I live? Where do I come from? Where am I going? And, many have asked without finding an answer to all these questions.

Philosophers, theologians, and strictly intellectual thinkers will not be able to respond correctly to these questions because they only manage ideas, beliefs and ideologies and do not base their studies on truths proven by Science.

---

[36] Motto of the Universidad Nacional Autonoma of Mexico engraved on its coat of arms.

Answers to all these questions regarding existence are in the laws of Nature and they can only be found through Science.

In the previous chapter we talked about the concept of the soul, but this entity is often confused with spirit.

As a background we can mention that the terms soul and spirit have been the object of many studies since time out of mind, and that up to now they have not been able to deduce a satisfactory difference between both terms.

The reason for this is that no scientist, philosopher, theologian or intellectual, has studied the manner in which the laws of Nature rule the bodies of all living creatures nor the way they govern human behavior.

This book has the intention to establish the difference existing between both entities.

I have found the first records talking about these two concepts in the Bible.

Genesis[37] mentions that the soul is the breath of life that God blows into humans and within said breath men receive the spirit also.

The Old Testament, the Pauline letters and the Gospel of Luke of the New Testament, mention on different occasions the soul and the spirit as two

---

[37] The Bible, supra, at note 14 .- Genesis: 2:7, 7:22.

different concepts but there is no way to identify comprehensibly said distinction.

Philosophers and thinkers of the ancient times considered that the soul or anime (from the Latin word '*anima*') is the beginning of life, of thought and of the sensations, that together with the body they constitute the essence of humans[38]. To them, spirit and rational soul were synonyms.

Descartes (17th Century) identified the soul with the mind.[39] He also refers to the soul as the vital force in animated bodies and in man as the spiritual incorporeal nature of mind or spirit. Plato, Descartes, and Berkeley hold that the soul is immortal.[40]

A few modern thinkers in the nineteenth c century contemplated the spirit as an immaterial principle that together with the body constitute the individual. Such principle was also attributed to all living things of the Universe (the *nous*) as a whole and even to inanimate objects, such as stars.

We owe Christian theologians the tridimensional representation of the individual: body, soul and spirit, but they do not clarify the difference between the last

---

[38] Filosofía Griega.- Origen de la Filosofía-Presocráticos-Sofistas y Sócrates.- Digital Version:
http://www.e-torredebabel.com/Historia-de-la-filosofia/Filosofiagriega/Presocraticos/Alma.htm

[39] René Descartes y el Legado del Dualismo Mente-Cuerpo.- Digital Citation: http://platea.pntic.mec.es/~macruz/mente/descartes/descartes.html

[40] André Lalande.- Vocabulario Técnico y Crítico de la Filosofía.- Editor: El Ateneo.- Buenos Aires. 1953.- Volume I, Pag. 48.

two concepts and since they do not define them, they use them ambiguously.

Now, I ask my readers:

Which parameters do you consider that should be used to establish the difference between the entities we know as soul and spirit?

In my opinion, in order to establish the difference between these two entities, we have to analyze all living creatures and define the abilities that are common to all creatures, and which abilities are exclusive of the human race which are not present in the rest of the living species, and then we can use that difference to determine the definition of soul and spirit.

In the previous chapter I mentioned that the soul is the energy that manages all the information that all living things have in order to work according to the program incorporated into their genetic code.

I also mentioned that the soul determines the exterior shape as well as the shape that each internal organ of the body of a living thing will have, which must be identical to the ones of their ancestors; it also manages the invariable functions of each one of the organs constantly renewing them and maintaining their original shape, handling both the management and healing processes after sustaining an injury.

Now, in addition to the abovementioned features managed by the soul, humans have additional abilities that differentiate them from other living species.

These abilities are, among others, spoken language, the capacity to produce ideas and comprehend them, to elaborate concepts, discern between good and bad, to reflect on their own selves, on their own sensations, on their own ideas, to love and want to realize an idea and feel the necessity for a Superior Entity.

In the same manner, humans have a sense of the moral in their activities, they seek longevity, take care of the environment, produce science and technology to provide consumables for their subsistence and for the subsistence of animals and plants.

These abilities or features, particular to the human race, which other living species lack are the ones I identify as spirit.

In other words, spirit is the entity that differentiates the human being from the rest of the living things.

I can define spirit as the energy that manages human behavior activating the features exclusive to humans such as love, conscience, morals, ethics, intelligence and feelings to be congruent with the laws of Nature. These qualities are reviewed at length in the following chapters.

Spirit is genetic, this is, it is universal for all human beings, and it exists as latent in the individual since its conception because it is a part of its subjective and immaterial structure.

In the previous chapter I made a very elemental analogy of the body of all living things with computers, comparing their body (hardware), their soul with the operative system (software) and life with electrical power.

Continuing with such comparison, spirit would be another program (or software) of humans containing exclusive features that I call spirit.

In the same manner I would like to tell my readers that by observation and reasoning I arrived at the conclusion that when humans reach full physical and physiological development, the mind is integrated into intelligence, memory, reasoning and conscience.

To me, the mind is the main and distinctive element of the spirit, and it is exclusive to humans. Its function is to propitiate the search for nourishment for the human society as well as comfort to achieve longevity.

In that same sense, I think that by intelligence and reasoning, the mind allows men to understand logic and the order present in the Universe, as well as the manner in which the laws of Nature affect us.

The mind receives information required to perform and order its functions through the central nervous system, through the senses, as the human's activities can be controlled at will.

The functions of the spirit are intellectual, immaterial, conscious, and subjective; they are expressed externally. These functions were mentioned in the previous paragraphs.

The soul, in turn, as we saw in the previous chapter, acts over the internal body of living things through their autonomic nervous system because its functions are unconscious, automatic and cannot be controlled by the individual.

The spirit does not act in the internal functions of the body (nutrition, digestion, respiration, metabolism, and other mentioned in the previous chapter) because these activities belong to the soul.

These activities, as I said in the previous chapter, are material, unconscious, and internal, this is, they are not exteriorized as the ones of the spirit and cannot be managed at will.

Living creatures in the higher taxonomic ranks have a memory and instinctive intelligence, but they lack introspective power that is a trait of the conscience, and of the capacity to reason and have ideas, as these qualities are exclusive to the human being because they are managed by the spirit.

The spirit is active in the human individual solely while he is conscious. The spirit stops acting when the human being is asleep or unconscious, but the soul is always acting and managing the inner (unconscious) body functions.

The search for longevity is a part of the survival instinct, and it is exclusive to the human being. It acts driving people to live for the longest time possible and to help others to achieve the same even when they are sick, handicapped or terminally ill.

In the second chapter of this book, I stated that the search for longevity managed by the spirit through the human will to try to live the longest time possible is different from the law of survival that all living things have.

The law of survival seeks only the longevity of the body already programmed in its genetic code when managing exclusively the inner physiological functions of all living things that are under the control of the soul, as nourishment, metabolism, digestion, etc. This longevity cannot be prolonged. It is what I call genetic longevity.

Meanwhile, the search for longevity in human beings is the motor driving the intelligence to prolong the genetic longevity and persuading it to produce food, medicines, and amenities to live as long as possible.

Under this pressure, the mind develops a great deal of abstract (subjective) ideas, and aided by reason, it selects those that will turn into Science and then technology for humans to provide food and amenities for subsistence and well-being.

Within the abstract ideas produced by the mind are those that materialize into buildings (schools, hospitals, highways, industry, etc.), music, poetry, painting, political and religious entities, means of transportation, forming families, dictating laws, manufacturing machinery and equipment, to produce consumer goods, technology, prescription drugs, etc.

Ideas are solely managed by the spirit; therefore, this ability is exclusive to humans; the soul manages the brain physiologically, this is, its organic functions (blood circulation, brain cell nourishment, etc.), but the spirit manages the ideas produced by the brain. Animals and plants cannot formulate ideas.

Humans may not know their future or the purpose of this life, but they are aware, due to the search for longevity, that there is a reason to live, that it must be for as long as possible and not to be defeated by death.

I think that due to sentimental reasons, or malnutrition, or indigestion of alcohol or drugs, an individual stops feeling the anxiety for the search of longevity and may make that anxiety worse and may become more anxious and depressed leading to physical decay and self-destruction.

This means that when the longevity functions are annulled, the survival instinct of the human individual disappears; thus losing interest in life.

To avoid this, human individuals must comply with the laws of morality described in chapter four of the second part of this book, to be emotionally and physically healthy.

Next I would like to ask my readers the following question: How do you think the spirit controls the instincts of a human being?

In order to answer that question we must remember the that chapter in which we talked about the laws of Nature we mentioned that basic needs come from such laws such as eating, drinking, sleeping or breathing; Nature gave the individual human instincts such as hunger, thirsts or sleepiness to comply with such laws.

I also mentioned that individuals tend to exceed scope of their instincts by eating, drinking or sleeping in excess due to the apparent, temporary satisfaction produced when fulfilling such impulses.

But in my opinion, it is the function of the spirit to control the excesses of said instincts to avoid displacement because the spirit is the moderator or regulator of all involuntary actions of the instincts by means of the laws of morality.

Now, to answer the previous question, the spirit is the one suggesting to the soul the activities that instincts should perform and not to exceed the fulfilment of their functions, even if the will of the human being is able to decide to allow or not allow the instincts to manifest in a way that fails to comply with the laws of Nature.

It is said that humans have two kinds of freedom: one coming from acting or properly abstaining from doing or thinking something and another one coming from free will, which is the ability that all human beings have to break the laws.

But this last ability cannot be called freedom. No human being has the ability or authority to infringe a law. This is, the capacity to harm someone or to violate a rule or have negative thoughts or abuse nutrition or sexuality. This cannot be called freedom; it's not freedom at all.

The limits of liberty on the individual in a human society are human laws and the respect for the rights of other members of the society.

On the other hand, I think that will, reasoning, love, feelings, character (personality), intelligence, morality, ethics, and conscience are genetic for all human beings. These are the main characteristics of the spirit.

The reason for this assertion is that those attributes are universal for all human beings; they are fully programmed in their genetic code and transmitted by inheritance.

But, they are discretional, or voluntary, and the acts that humans voluntarily do include:

- Deciding and governing their conduct (behavior);

- Their thoughts arising from the ability to reason;

- The processes of feelings and intelligence they have;

- The deeds of their temper (attitude, mood manifestations);

- The deeds of their conscience.

These acts are not transmitted by inheritance, they differ for each human individual, and therefore they are not laws of Nature.

It is necessary to clarify that when humans manifest their thoughts or deeds coming from free will they may exceed their genetic programming and chose to comply with a law or omit such law and perform an activity or fail to do it, have positive or negative thoughts, accept this idea or reject another, etc.

But if these thoughts, feelings, deeds or emotions exceed their genetic programing and fail to comply with a law of Nature, the body shall sanction the individual causing an imbalance that will evidently reduce longevity.

Now comes to mind a question that has concerned human beings since ancient times:

What is the meaning of human life?

From an objective point of view (excluding any philosophical, metaphysical or religious thought that cannot be proven) I consider that the only thing that can be demonstrated is that all activities performed by human beings in their lives have two purposes, survival and longevity.

Survival includes the conservation of the species while longevity includes the amenities and relaxation searching for the welfare of the individual.

Then, if an individual works, he does it to for profit to get food, this is, to survive.

Income is also used for recreation (traveling, movies, reading, dancing, music, sports, religion, art, meditation, social gatherings, etc.), which activities reduce stress and achieve a better longevity because they are comforting.

And, if the individual acquires goods providing amenities or satisfaction (comfortable beds, automobiles, electronic equipment, etc.) these will help him to achieve longevity.

In the same manner, if the individual performs altruistic activities, that is love, love favors homeostasis, harmony or inner equilibrium producing longevity.

Additionally, if humans have sexual relations, they may do it for simple pleasure, in which case it would be a calmative, or it might be that they do so in obedience to the commandment of reproduction that is a part of the survival of the human species.

We can say the same thing if the individual takes care of the environment because it benefits his survivorship.

Better yet, what we said in the previous paragraphs regarding the meaning of human life may be reduced to just one thought: **The purpose of human life is to live in harmony with the laws of Nature.**

From all of this we can conclude that **spirit is the energy driving humans to preserve the Creation and to create the conditions to produce goods necessary for life (survival) and comfort (longevity).**

Some people confuse spirituality with religion, or with feelings or meditation or with something esoteric or intangible because they are unaware of what spirituality means or they have an ambiguous perception.

In my opinion, spirituality consists in abiding by the laws governing love, moral, ethics and the feelings that are the laws of the spirit.

This means that spirit is the energy that vivifies the laws of Creation, exclusive to the human race.

Humans are in harmony and equilibrium when the soul acts in accord with the spirit. Man is only one; soul and spirit in one body.

When humans, by their own will, make their instincts obey the laws of the spirit, this individual is in harmony and equilibrium; he is in peace.

Peace is the most appreciated asset of human beings, and it is the foundation of happiness.

But, who are the immortals for you dear reader?

For me, the spirit of each person is immortal and once it becomes part of the universal spirit, it remains in history and in the thought of all human society and of all the peoples and times of the world.

Personages that roamed the world leaving their work for the benefit of Humanity are called by society immortals (Beethoven, Gandhi, Edison, Pasteur, etc.).

Chapter II

Instincts

Instincts are numerous and they play an important role in the functions of the organisms, but we shall only study those influencing the behavior and conduct of the human individual in his contact with the outside world, society and the existential processes in which individuals participate.

In my opinion, **instincts are natural and involuntary impulses, not experienced or learnt, that respond to inner stimuli (that may be imaginary) or external, and are for the individual to seek his survival and reproduction.**

The laws of Nature act in living beings through their instincts.

Instincts move humans to act without thinking first and before pondering actions and their impulses are obeyed regardless of conscience and reason. The latter is the main characteristic of instincts.

The main function of these impulses is for the bodies of all living creatures to work properly in accordance with the laws of Nature programed en each body.

Instincts are engraved in the genetic code of each organism, and they are typical for each gender and member of the same species because they are transmitted by inheritance and are managed by the soul of that living thing.

We said on many occasions during the development of this book that the main instinct of all living creatures is survival and in order to comply with such demand, the laws of Nature order eating, drinking, reproducing, sleeping, etc. and through such instincts, living creatures are driven to comply with these laws by feeling hungry, thirsty, sexual desire, sleepy, etc., making it possible for the individual to feel pleasure when matching said satisfiers.

Now, no living thing –with the exception of human beings– exceed the satisfaction of their survival instincts because they are programmed in their genetic code, unless such creatures are captive so that its equilibrium with Nature is broken.

The possibility to control such instincts constitutes an important difference between animals and humans.

Humans must control their actions from instincts in order not to exceed and to prevent damaging the body;

animals, in turn, in their natural habitat, obey said instincts blindly.

To be clearer: all living things (exception again ade for humans) eat only when it is necessary to survive even when they have excess food at the time they are eating.

Therefore, natural or instinctive human impulses tend to exceed the requirements for food, beverages, sex, sleep, etc., because they feel satisfied when they do, but they must control them at will to prevent damaging their health.

The laws of morality govern instincts, limiting and moderating their excesses. This is better explained in the chapter dedicated to morality.

The laws that govern the functioning of the senses, such as of the eye, ear, nose, etc., are universal in all humans and should not be confused with the instincts of vision, hearing, perception of odors, etc., which are different in each person.

The normal and extraordinary abilities that a person acquires or develops such as those needed to sing, run, play a musical instrument, etc., are particular to each person and are not transmitted by inheritance.

Humans have the capacity to neglect or disregard her instincts or reduce their spontaneous manifestations postponing their compliance even if they cannot be fully suppressed.

Following the foregoing, it is appropriate to clarify that, initially, instincts act spontaneously regardless of the will and the reason of the human individual to seek survival.

But later, initial impulses, thoughts, acts or actions executed by human beings to satisfy their instincts belong to the temperament and will of the individual; therefore, individuals may exceed their compliance, but they can control them also because these thoughts, deeds or actions are not genetic but acts of volition.

Instincts towards what is sacred or religious are exclusive to humans and must be satisfied individually, particularly for each human individual in the same manner as the instincts of hunger, thirst, etc., because instincts are not the same for each individual.

Every human has his own particular way to express feelings towards religious matters because it is a reaction of their temperament and of the feelings of each individual and nobody can make other individuals have or express religious feelings in a determined manner.

Instincts look for survival and moral for longevity. Laws of morality, through intelligence and the will of the individual, must control excesses in instincts in human beings so that their conduct complies with the orders of the laws of Nature.

Chapter III

The Universal Conscience and the Human Conscience

"*I find my rules of conduct* in *my heart. Whatever I feel to be good is good. Whatever I feel to be evil is evil. Conscience is the best of casuists.*"[41]

- J. J. Rousseau.

To better understand it, is appropriate to establish the difference between the terms "universal conscience" and "human conscience" to avoid different interpretations.

Universal conscience is the ability that all human beings have to be aware of the environment surrounding them by means of their senses and to create actions according to said knowledge to procure survival and comfort.

---

[41] Frases Célebres de Juan Jacobo Rousseau.- Digital Version: http://www.elasere.com/interes/ShowFrases.asp?id=110&Letra=J&exist=no

In addition to the abilities given by the universal conscience, humans possess the quality to reflect and to distinguish between right and wrong and of self-discovery.

These abilities distinguish humans from other living creatures and this is what I call human conscience or reflective conscience, which is a characteristic of the spirit.

Animals lack reflective conscience, and they lack moral (the capacity to judge the goodness or evil of their actions).

Reflective conscience is an element of human Nature. It is an integral part of a human's subjective structure and is immaterial (like the soul, life, ideas, mind, etc.). These properties are innate, universal in all humans, and not acquired.

The study of conscience has walked a long road in human history. I can assure that humans, from the time they were able to express ideas and knowledge by means of writing started to study, both universal and human conscience, as well as their respective functions, even though often in a confused way.

Conscience is infused in the human body since birth, but it remains latent, and it is developed during childhood reaching its maximum capacity when the individual enters puberty.

In the same manner, **I can assert that conscience is the entity in which the laws of Nature affecting exclusively human beings are recorded. I call them laws of morality for the reasons I give in the following chapter.**

Actually, conscience has been experimentally identified and its functions proven both systemically and statistically.

It is an experimental fact (only "experimental" so far because conclusive proof is still lacking) that disorders appear in the metabolism of those individuals failing to comply with a law of Nature, they are produced by the actions of the internally secreting glands.

I think that human conscience is an indicator of the transgressions committed against the laws of Nature. This entity monitors and alerts the actions and thoughts of the individual so his conduct is congruent with these laws.

But no indicator is infallible because its directions change for each individual, and additionally, in many cases, conscience can be subjected to the person's volition preventing the individual from realizing the breaking of a law of Nature.

It is not a judge of human actions either; it only warns the individual when one of his deeds is breaking a law of Nature; that is, it merely records thoughts, feelings, emotions, and actions of the human beings and tells the person if his or her actions are congruent with those laws.

As I see it, I think conscience is like a control center or monitor recorded in the "perfect" model of the human being, an ideal image known as the "inner self" or "super ego" as it is called by psychoanalysts.

According to the ideal human being image incorporated into the conscience, internal secretion glands secrete well-balanced biochemical liquids making the body to work in harmony (homeostasis).

When humans perform something that breaks a law of Nature, the brain registers said action and sends an indication to the conscience and the conscience then detects the infraction because it fails to match the ideal human image it has already recorded, sending an alert to the individual as a kind of "inner voice" notifying the infraction, after that, internal secretion glands receive an order to produce biochemical liquids in excess over normal levels produced by the body and discharges them into the blood.

These secretions unbalance the offenders' etabolic functions causing damage to their bodies.

We have no way of knowing the magnitude of the damages caused by the infringement of a law of Nature because we also ignore, in most cases, the laws governing human behavior.

Therefore, what I can really affirm is that not conforming to any law of Nature, even if they are ignored by the infringer, harms that individual's body reducing longevity or receiving other unknown natural sanctions.

Now, how can an individual know if he is breaking a law of Nature if he is not aware of most of them?

I think that the best way to detect an infraction of any law is when a person acts in a secretive way or tries to hide an action.

This means that the individual does not want anybody to know that he is doing something wrong. Everything done covertly is wrong even if it does not harm others. The person is aware that he is doing something wrong.

Now I ask my readers, would you accept for pleasure or a substantial amount of money that someone take a picture of your genitals, without anybody knowing whom they belong to and without showing the face or any other recognizable part of your body?

In this case, I can also say that any action performed by a person to induce another to break a law of Nature is against the laws of morality.

Conscience and morality are intimately linked to each other in such a manner that the relationship between them is direct, and I can say that one could not be defined without the existence of the other as we will see in the next chapter.

Chapter IV

Moral

"The most perfect morality" shall not proceed but when all laws of Nature are known, because that knowledge is the outmost grade of wisdom [42]

- Descartes.

I said in the previous chapter that the purpose of conscience is mainly to register the laws of Nature particularly affecting human beings.

I call these the laws of morality. We should not call them moral laws because we would have also immoral laws. Immorality is not a law, but a non-compliance with the laws of morality.

I also said that conscience and morality are intimately linked and that the relationship existing

---

[42] Regis Jolivet.- *Tratado de Filosofía. Moral.*- Editor Lohle.- Buenos Aires, Argentina.- Pag. 194.

between them is direct and one cannot be defined without the existence of the other.

In my opinion, **the laws of morality are the group of rules governing human behavior as per their nourishment, sexuality and feelings by suggesting the individual moderate excesses in order to maintain the homeostasis (equilibrium) of his body can achieve its maximum longevity.**

These laws act directly on the survival and conservation of the species instincts besides longevity.

This group of laws is incorporated into the genetic part of the instincts of each human being, and they make that the internally secreting glands work in optimal condition to achieve the best health and maximum longevity.

I will try to explain myself: I have said on many occasions in this book that the laws of Nature command eating, drinking, reproducing, sleeping, etc., in order for humans to survive, and by their instincts, driving humans to comply with these laws making the individual feel hungry, thirsty, and have sexual desire, sleepiness, etc.

But instincts tent to exceed the achievement of these satisfiers because the individual feels pleasure when doing so, but the laws of morality suggest the individual limit and moderate such excesses in order to have physical and emotional health.

This means that by using one's own intelligence and volition, humans must control such excesses in order to obey the laws of morality.

This is what is known as temperance, and it is the main attribute of the laws of morality.

I said at the beginning of this chapter that the laws of morality regulate the nourishment, sexuality and the feelings of the human individuals. We are going to study each one of these elements:

In the first place and relevant to all the laws governing human nourishment, I am able to say that human body is programmed to work in optimal conditions with well-balanced amounts of food to survive.

In this manner, the laws regulating nourishment, suggest the soul (the body) eat in moderation, in a balanced manner and only the food required by the body to survive, work in optimal conditions (carbohydrates, proteins, fats, vitamins, minerals, etc.) according to the status, age or physical condition of each individual.

But human volition, that is, free will, may refuse to hear such call and exceed those requirements eating in excess due to the pleasure provided by satisfying the hunger instinct (food) affecting the health of such individual. We ignore how much, but certainly, the damage will be serious.

In fact, as I said in the previous chapter, humans are programmed to live one hundred and twenty years, as scientists were able to prove scientifically, additionally, by analyzing objectively this assertion, the fact that people arrive at such age means that we all are able to do so since all humans have been created equal to the laws of Nature.

Nonetheless, at the present, humans live and average of seventy to eighty years, which means that individuals have been breaking what these laws order, and this is why the life term for which they were created cannot be reached.

Nutritionists have the task to determine the amount and quality of food and the schedule to be followed by each person according to their age, activity and state of health. If an individual eats irregularly or in excess to the amounts established by the law in its genetic code, that person is acting immorally and at the same time, in detriment of his health.

In the same manner, besides the fact that humans must eat and drink whatever is proper for their bodies, Nature obligates people to breathe pure air. If people breathe impure air or take beverages unnecessary for the body, individuals may survive but longevity shall be reduced.

Humans may be taking adequate and well-balanced amounts of food but experience health problems anyway. One of the causes could be that the individual is breaking another law of morality as we will see further along; sexuality and passions negative such as hate, resentment, anxiety, stress, depression among others.

Second, sexuality is another law of morality that manages the conservation of the species. Humans are gifted with a sexual capacity sufficient to procreate descendants and comply with the reproduction commandment.

This sexual capacity is subject to the laws regulating the use and frequency of their sexual activities, including their thoughts to reach the maximum longevity in addition to the conservation of the species.

There are no scientific studies regarding these laws like the ones regarding food that are being studied more deeply; nevertheless, as it occurs with all laws of Nature, abusing sexuality damages the body or be the cause of other natural punishment we do not know.

I shall explain this: zinc is an essential element for the human health. It is indispensable for the correct function of the nervous system prostatic gland and brain cells. It is necessary also for the correct function of the optical system: photophobia or extreme sensibility to light is caused by a zinc deficiency.

Human semen contains multiple minerals denoting a high zinc content.[43][44] It is estimated that in each ejaculation a milligram of zinc is eliminated, therefore if a man abuses sexual activity he is exposed to health deterioration due to the lack of this element.

I think that when young people abuse their sexuality (including masturbation) they are exposed to debilitate their optical system and their brain cells and as adults their prostatic gland as well.

I can assure without a doubt, that there is cause for everything in Nature; acts of Nature are not unexpected; Nature always warns the individual when he is about to be injured when exceeding the laws the laws of morality regulating sexuality.

The problem is that individuals neglects the warnings given by their bodies or they learn to live with the symptoms of such warning.

I think that in some cases, erectile dysfunction –when not caused by emotional or health issues– is a call of the Nature to an individual warning that person that he had exceeded in sexual activity, and it is time for a break.

If the individual disregards this warning and continues abusing his sexuality, it shall cause

---

[43] Liz Hodgkinson.- *El Sexo No Es Obligatorio*.- Javier Vergara Editor, S.A.- España.- 1988. Pág 73.
[44] Discovery Salud.- Digital Version:
http://www.dsalud.com/index.php?pagina=articulo&c=390

irreparable damage to his body, and in some cases it may incapacitate or cause death.

Basing my opinion in observation and reason although I have no proof, in some cases, injuries such as brain strokes are the result of a person's excessive sexual activity.

On the other hand, I think that abusing sexuality has been one of the main causes of the problems for humanity, because, besides it has caused overpopulation in the Planet, thus deriving into lack of water, food, as well as the pollution of rivers, lakes, seas and the environment.

Third, the laws of morality govern human feelings through their code of love regulating the passions negative, causing the internally secreting glands to function at its best, thus aiding the individual to live healthier and achieve maximum longevity as well as the longevity of others, and thus increasing the life expectancy of all human beings.

The law of love is constituted by the total of positive feelings of affection, endearment, kindness, patience, forgiveness, charity, and others, as we will see in the next chapter dedicated to love.

Humans are designed to live in homoeostasis, in love, and in inner and outer stability. These are the best conditions of the human body to attain health and greater longevity.

In the same manner, positive feelings are called that way because they favor metabolic equilibrium making the individual healthy, and when this is achieved, individuals live in a state of benevolence, of love.

I call negative feelings passions negative, and they are the antithesis of love. Passions such as hate, resentment, jealousy, vanity, envy, lust, obsession among others, are self-destructive.

Meanwhile, by having passions such as selfishness, wrath, vengeance, greed, among others, the individual damages the society where he lives, in addition to the self-inflicted damage because they disrupt human equilibrium (peace).

These negative feelings, particularly selfishness, silence the law of love and hinder a human's ability to correctly reason, and when they cannot be controlled by the individual, they provoke social violence.

On the other hand, humans are also provided with emotions that may be positive or negative. Positive emotions such as joy, euphoria or enthusiasm, among others, benefit longevity.

Meanwhile, negative emotions such as stress, anxiety, anguish,[45] depression, sadness, fear and others, provoke excessive hormone secretion provoking

---

[45] Oppressive fear without precise cause. – Pain or suffering. – State of extreme upset before an undefined danger or vague, unexplained or indeterminate threat.

metabolic unbalance and illnesses that reduce longevity and jeopardizing her survival.

In the same manner I sustain that moral principles establish self-duties for the human being; individuals must fulfill those principles completely, without any excess or deficiency.

To the laws of Nature, humans have no rights, only obligations and failure to comply or any infraction shall be suffered by the offender's body.

If a human individual commits an infraction against a law of moral, it shall be known only to such individual. Third parties are unable to judge if the act is moral or immoral.

The compliance or infraction of moral principles shall determine the degree of morality or immorality in which the active subject may be rated, but only said subject would be able to determine if it was moral or immoral.

Nobody outside the actor may judge the quality of his actions because he alone shall respond for actions to one's own conscience, that is, to the inner-self.

Because the laws of morality are inner laws, they set forth obligations to oneself. Nobody can compel another to fulfill a moral obligation because the fulfilment of these laws are personal and inner obligations.

Nevertheless, when immoral conduct is exteriorized (such as hate, resentment, vengeance, wrath, ambition, etc.) then it enters the field of ethics, and if these actions damage any member of society, it may be sanctioned by the latter, in addition to the damage caused to oneself and to the body. We shall see this in the following chapter.

I think that good and evil in human societies are motivated by the actions of our instincts. Evil (the devil or Satan) is the actions of instincts that overflowed the laws of Nature, good is the conformity to those rules.

In my opinion, the human individual is moral when he complies spontaneously, unprompted, with the laws of Nature.

On the other hand, individuals are immoral when they consciously infringe the laws of Nature, without coercion and before their conscience as a witness.

Failure to comply with the laws of Nature due to ignorance or coercion does not mean that an individual is immoral; however, the punishment arising from the transgression of the law is unavoidable.

The course taken by morality through the years inclines to the divine and religious. Religions have taken possession of morality and converted it into rectors and caretakers of the conduct of their congregations and proselytes; nevertheless, all human beings are subject to the laws of moral, regardless of the religion or ideology they profess.

I consider necessary that schools teach the laws of morality to give children since they are small, giving them the knowledge they require about the laws of nourishment, sexuality and human feelings, including love for one another and for Nature.

Children should also learn the negative consequences provoked by the disobedience to any of the laws, so they may live a healthy life, both physically as emotionally, and be able to live in peace in their society.

I make this recommendation because the laws of morality are eternal and invariable, and they are applicable to all human beings from every country without distinction of religion, race, gender or social status.

Great philosophers and scientists that have lived in this world have only studied the laws of Nature governing physics and chemistry, that is, the laws they were able to weigh and measure.

But if they had studied the laws of Nature that govern the conduct and behavior of the human beings, the world would be another thing, very different from the one we have now.

For this reason, I invite my reader to try to discover the laws of Nature that govern human beings and those that apply in his or her own benefit to improve the quality of life as well as that of the environment and of human society.

On the other hand, I want to tell my readers that some thinkers believe that morality originates in customs and traditions, and therefore, it has to be updated with the passing of time or a change in the mentality of each person and that society is the one deciding what is moral or immoral in one determined era or place.

This thought is wrong because the differences that exist between morality and customs or traditions are totally opposed.

Said differences, to my judgment, are the following:

1st. Morality is eternal and invariable; it does not change with the passing of time. Customs or traditions, in turn, are temporary and variable according to the place they are practiced.

2nd. Morality is universal, that is, identical for all human beings, and in any part of the world, because it is genetic. Customs or traditions are local, they are not genetic, but acquired, and they change constantly.

3rd. Morality is internal and individual, that is, each human must comply internally and particularly with his moral obligations. Customs or traditions, in the other hand, are external and plural, because they are exteriorized by human beings in the community where they live, and,

4th. Morality is a law of Nature and creates internal obligations for the individuals. While customs or traditions are not internally obligatory, but they are external for the individuals.

On the other hand, we can observe people with different customs, and we will be able to demonstrate that the ones complying with the laws of morality have a longer life expectancy than those that fail to do so. This evidences the difference between both concepts and the validity of the laws of morality.

Chapter V

Ethics

By this all people will know that you are my disciples, if you have love for one another [46]

- John 13:35

Greek philosophers, Socrates, Plato and Aristotle dedicated many of their studies to ethics. In that era, history told us that the ethics and moral concepts were synonyms.

The difference between both words lies in their origin: *Ethos or ethĭkòs* meant ethics or morality for the Greeks, and it meant custom or habit too. *Morālis* meant morals or ethics for Latin people and meant character too.

Modern dictionaries give those words said synonym identity; nevertheless, the use of one word

---

[46] The Bible, supra, at note 14.- John 13:35.

or the other does not always have the same meaning in literature, philosophy and judicature. Their meanings in the colloquial language differ also, and they are not used as synonyms.

For example, the Socratic school considered that ethics is the virtue to live according to the laws of Nature.

In the seventeenth century, philosophers thought that ethics should have a moral sense that makes people act according to what reason ordered as good or bad.

Since that time, a difference existed between the meaning of one word and the other. In the nineteenth century, it was said that man was not naturally moral, but ethics should convert him into a moral being.

Even though Aristotle knew ethics and knew the existence of the laws of Nature, he could never link them with the moral, because he was not aware that these laws also ruled human behavior and influenced their happiness.

Unfortunately, this way of thinking is still in force –even now, because the definitions of ethics and moral given by modern dictionaries are ambiguous. There are no precise definitions; therefore they are used freely by writers, and that is why I proposed to myself in this book to establish, according to my criteria, the difference between these terms.

I shall define the word ethics related to the moral action. Ethics and morality are a part of love, as we shall see in the next chapter.

Ethics is the spontaneous exteriorization, by free will and without external coercion of the moral conduct of the human being before the society in which we live.

Ethics, in certain way, is a justice equivalent of the moral individual when this action is exteriorized and it affects others.

For me, ethics are the correct actions of a human being. To judge if those actions are correct or not, one should go by universal parameters, in other words, those that all for all individuals regardless of sex, race or nationality. Only the universal ones are given by the law of Nature.

This is a summary of behavior I call ethical of human beings:

- Human beings are responsible for the survival and longevity of their fellow-men;

- They love their neighbors even enemies that may harm them;

- They act with selflessness and do not expect any retribution or compensation for their actions;

- They are fans of the truth: they do not lie; sincerity and righteousness are their invariable norms;

- Their faithfulness is impeccable;

- They do not avoid their obligations;

- They do not respond to harm with harm;

- Humans are true to their promises;

- They never take advantage from their business dealings;

- Humans obey the laws derived from society;

- They do not abuse the good entrusted to them by others;

- They respect the dignity of others (their human rights);

- Do not breach the laws of Nature;

- They do not obstruct nor interfere with natural processes;

- Humans respect ecological equilibrium and replenish it when it has been broken;

- They try to avoid polluting the rivers, the lakes, the seas, and the environment.

In my opinion, these human activities constitute ethics. **Ethics is the common sense that all human beings** should have when helping others achieve their maximum longevity.

An external conduct is ethical when it agrees with the laws of morality.

Ethical value disappears when human beings obtain pleasure while acting.

Ethics demands putting good ideas in practice; good feelings have no ethical value, just moral value and they only benefit the individual not the society. Good feelings belong to the moral field.

The difference between morality and ethics is substantial because the moral is internal and unilateral; whereas, ethics is external and plurilateral.

Moral refers to personal inner judgment while ethics refers to social values; compliance with ethical and moral values must be spontaneous and altruistic.

The opinions and consensus of human regarding ethical principles do not modify the universal validity of those principles; no democracy in the laws of Nature exists.

Currently, ethics and morality are not frequently used as synonyms because many literary, philosophical, juridical and other writings use the term ethics with a given meaning, and if we try to change said phrase with the word "moral", the sense of such expression would change.

The difference between ethics and morality expressed in the previous paragraphs is categorically evidenced when an individual is personally immoral (in his thoughts), but acts ethically in his business transactions.

Professional ethics may be compared with honesty and loyalty because it refers to the relationship of an individual with third parties whether the thoughts of such individual are moral or immoral.

This is why professional ethics may not be called professional morality.

When the deeds of an acting individual are not prejudicial nor cause harm unto others, but internally fail to comply with the laws of morality, the actions of such individual are immoral before his own conscience but no one may demand obligations or claim anything against him because said action was merely internal, moral (or immoral, in this case).

But when human actions are exteriorized to harm others or harm Nature the individual is acting without ethics or being unjust and they may be punished by society because compliance with the ethical principles is subject to the judgment and approval of society.

This means that when an individual's thoughts and deeds are just, that individual will be acting morally and ethically correct unto others.

I can summarize with the following words: **Ethics is respect for Nature including for other human beings.**

A clear example regarding an action lacking ethics is the case of an individual that refuses to help a hungry person when her position and situation allow her to do so.

The individual, who is refusing to help, is acting unethically, because human beings have the obligation to contribute to the survival and welfare of others; however, nobody can call her immoral because she cannot be forced to help others.

Chapter VI

Love

Love, even though human, is divine... because even God loved.[47]

- Felipe Pinglo Alva.

I can define love as a law of Nature that forces the human being to surrender selflessly to the service of others with intelligence, will, and energy and to the conservation and improvement of the environment.

I sustain that love is a law of Nature because it is common (universal) for all human beings.

Love is the respect of the laws of Creation, of all that was created, and it is manifest in attitudes, words and deeds.

---

[47] Popular song.

Love is constituted by morality and ethics, internal and external behavior, respectively in all human beings.

Christians sustain, over all, the primacy of charity over justice. Pope John Paul II says, "Love triumphs over justice and it complements it following the logic of surrender and forgiveness."[48]

Love is formed by the sum of all positive feelings, such as affection, endearment, tenderness, friendship, humbleness, charity, patience, forgiveness, altruism, peace, tolerance, generosity, meekness, mercy, and many more.

We cannot individually call all these feelings love because they are only an expression of love, but together they constitute love.

When humans integrate love into their daily ways, thinking and living, their bodies work in harmony, benefiting their health and longevity.

Ashley Motagu[49] tells us that humans are born with an innate need for love, with a need to respond to it, to be good, cooperative. Everything opposing love, goodness and cooperation is inharmonious, chimerical, unstable, and dysfunctional; it is bad.

---

[48] Juan Pablo II.- Encíclica *Dives in misericordia*.- Digital Version: http://www.instituto-social-leonxiii.org/index.php/estudios/844-la-animacion-de-la-caridad-en-la-comunidad

[49] Ashley Montagu.- *¿Qué es el Hombre?* (2ª. Ed).- Editorial Paidós Ibérica.- 1987. Pag. 101.

Montagu continues by saying that if the needs of a child were adequately satisfied, the child could not be other than good, that is, loving. All natural inclinations of men are directed to the development of good and towards the interruption of the displeasure states.

I consider that there is only one kind of love: filial, maternal, platonic, spiritual, and "sexual love" is merely rhetorical.

Inappropriately, sexual intercourse is called love. This mistake is made because the Greek language includes the words *agape* to designate love and *Eros* to designate sex. But the problem of the ambiguity of the term love in other languages appears when translating Greek to Latin. In Latin there is only one word for both Greek meanings of love. Both English and Spanish, like Latin, drags in ambivalence.

Pleasure means a gratification that is given by God to the human species to force them to comply with the reproduction commandment to keep up the species. While complimental to the satisfaction of the sexual pleasure, affection and fondness intervene, it is only then when we can call it love, but it is because of these feelings and not due to the sexual act.

Love must be permanent at any age. Desire is transitory at any age. Love is a permanent uniting element. Sex separates more that it unites when performed without love.

Sex is not required at all when we are giving love to others, but sex does respect the laws of Creation, therefore it is in fact a type of "love".

It is important to mention that humans have only two unique and indispensable pleasures for the survival of the human race, food and sex. All other satisfiers, those ill-called, such as the pleasure to travel, read, listen to music, and others may not be called pleasures because they do not reach such degree or the intensity of the former two, but also mainly because they are not universal, this is, not all human beings have them nor are they transmitted by inheritance, and therefore they are not laws of Nature.

Do you think dear reader that animals have the capacity to love?

I think that love is a manifestation of the spirit, thus it is exclusive of the human race. Animals do not have the capacity to love; their vital functions are genetically programmed.

Nevertheless, some animals show hints (appearances) of love expressed with tenderness, fondness, and affection, but they lack the rest of the positive feelings that constitute love and that we mentioned above (charity, altruism, generosity, mercy and others). Hints of love in species inferior to men are only that, hints.

Humans are designed to survive in harmony, in love, in inner and outer equilibrium, even when there is evil. Positive feelings are called that way because they favor homeostasis or metabolic equilibrium and character harmony (way of being). These feelings create the health of the individual and consequently, they favor longevity.

Passions negative are the antithesis of love; passions are constituted by negative feelings such as hate, resentment, jealousy, selfishness, ambition, anger, greed, envy, anxiety, anguish, stress, obsession, vanity, wrath, lust and other.

People who liberate their passions negative secrete chemical substances exceeding normal levels thus unbalancing their metabolic functions.

This of course, produces illnesses or lesions to their bodies reducing their longevity and endangering their survival.

All humans are born with the ability to be abusive, arbitrary, take advantage of others, selfish, dishonest, corrupt, etc., and some exteriorize said actions using their free will, but the laws of Nature (love, moral and ethics) demand that individuals must dominate said passions for their own benefit and for the benefit of others to have health and at the same time to achieve social peace.

People that fail to control their passions and cause harm to others, are predisposed to be sanctioned by the authorities of the human society, and if that does not exist, "justice" shall be exerted by those who think they have the power to do so.

Now, I have a new question to ask my readers, if passions negative are a part of human Nature, and they harm the human body, what is their purpose?

Evidently, passions negative are a part of the human nature of the human being, but I consider that Nature has given the human body with all those negative feelings to be used in case of emergency when the individual needs to defend his survival, but they are not to be used for a long time because they damage the human body.

As an example, if a lion attacks a person, said person may fight or run, but if the person constantly lives as if the lion were attacking him, the consequences of such altered state shall damage the health of such individual.

In that same order of ideas, I want to tell my readers that some people consider God as the creator of good and evil in the world; nevertheless, this book shows that the Creator of the Universe is not the author of what humans consider "evil" because evil is caused when humans exceed in their exteriorization of passion.

The sole antidote against negative feelings in human beings is living with love. When all humans live love with integrity, society shall have peace.

Love must be a permanent experience in the life of all human beings. Love necessarily implies forgiveness of offenders because that is a part of love.

Living a life of love is not only having positive thoughts of good will, peace and harmony, but also doing the good as well.

It is opportune to ask my readers, what does happiness mean to you?

Since ancient times, and even currently, the search for happiness has been one of the main concerns of humanity; many books have been written trying to teach people a way of living to be happy. They have gone so far as to assert that happiness is the objective of human life.

But happiness, as it is commonly conceived by the people is an undetermined and variable asset, and nobody can retain it because it is provisional and transitory; it is a state of mind.

Happiness is intangible, and it is expressed as a feeling of joy and contentment by whomever receives or takes possession of what he considers an asset or a satisfier.

Each individual has his own idea of happiness; therefore, I dare assert that it is not universal but a product of human feelings, and thus it is different for each one.

As an example, for a hungry individual happiness could be found in satisfying his appetite. For the mystique or religious, happiness could mean living as a recluse in a monastery; the greedy individual may find happiness in money; womanizers might find happiness in "possessing" more than one woman, and for others it may mean to achieve a goal. All these objectives are momentary.

Notwithstanding the foregoing, if humans live in harmony with the laws of love and morality, they shall have bodily and emotional health that could be deemed a foundation of a stable and possible happiness.

Nevertheless, if an individual fails to live under those conditions of harmony there are systems and methods of relaxation and meditation to find and help achieve it then he will be able to find what is most desired by men and women, peace, understood as the sentimental, physiological and psychological equilibrium of the human being.

Peace is the biggest asset that a man can obtain; it is universal, and it may be permanent, in human terms, and it is genetic and natural; it is what we might call true or universal happiness, but it is better to call it by its own name, peace.

Humans have been created to live in love. Peace is a part of love, and it is the only way to happiness for human beings. Peace is the purpose of Humanity.

Chapter VII

Homosexuality

Homosexuality may be as old as human kind. The word homosexuality comes from the Greek root *homo* (equal), meaning sexuality between equals, same sex; applied both to men and women.

When the word homosexual transcended to the Latin towns, the prefix *homo* from the Greek root was confused with the Latin prefix *homo* meaning "man", therefore homosexual was used to designate men having relations with other men; to differentiate the relationship between women, they called homosexual women "lesbians."

The word lesbian originated as patronymic from the Greek Island named Lesbos known because women living there rejected sexual contact with men as they considered men inferior.

Nevertheless, who first used and made popular the term Lesbia was the Latin poet Gaius Valerius Catullus a contemporary of Gaius Augustus Cesar, Caligula. Catullus called his lover Claudia, Lesbia –a Roman aristocratic woman, very influential and known for her scandals and libertine habits and to whom he dedicated his most burning and passionate poems.

Any reliable antecedent of the homosexual practice in primitive tribes comes from prehistoric era, and there is reliable evidence of those homosexual behaviors in human tribes and congregations in centuries prior to the present.

The Old Testament (Genesis 16:1-20) narrates the destruction of Sodom and Gomorrah because of the moral degradation in which their inhabitants had fallen.[50] Even though the defenders of homosexuality allege that they were not destroyed because of the homosexual practice, the fact is that the word "sodomy" has stayed as a synonym of homosexuality.

In some ancient civilized nations, homosexuality was practiced and accepted or tolerated as in Greece and Rome. But among contemporary primitive peoples it was generally prohibited and restricted, and it was punished with the expulsion of homosexuals from the tribal society or they were punished even with the death penalty.

In the Old Testament (Leviticus 20:13) it was prohibited, and it was penalized by death: "If a man also lie with mankind, as he lieth with a woman,

---

[50] The Bible, supra, at note 14- Genesis 16:1-20.

both of them have committed an abomination: they shall surely be put to death; their blood shall be upon them."[51]

Civilized people, during the ninteenty century and the beginning of the twentieth century homosexuality was practiced underground, hidden, and was not accepted socially by morally educated people, but it existed and it was propagated notwithstanding its rejection.

Nowadays and increasingly after the World War II with the liberation of the traditions and the development of human rights, homosexual groups have gone further to claim their equality in legal rights and social and labor, fighting non-discrimination.

A lot has been discussed in the present regarding the causes and social effects of homosexuality, and there are great discrepancies in the opinions to that respect; if said sexual practice should be accepted by human society as the individual right that each individual has to express freely privately and without outside limitation, because as it is alleged, said conduct does not harm any third party and homosexuals, as each individual, have the right to use their body as they deem convenient.

Some intellectuals, men of letters, philosophers, writers, poets, musicians, etc., and other bankers, entrepreneurs, sportspeople, scientists and even

---

[51] The Bible, supra, at note 14.- Leviticus 20:13.

common individuals are convinced and take as a fact that the origin of homosexuality is genetic and hereditary. This is fallacy.

Homosexuality means when two male individuals have a sexual relationship between them and not when they have a friendship relationship.

Male individuals with feminine features are not homosexuals if they do not exercise or practice sexual acts.

Many examples exist in all human societies whereby the many individuals with feminine features remain celibate and they may not be called homosexuals.

Two men with feminine features are only homosexual when they have intercourse with each other and not when the only have a friendship relation.

This means that **homosexuality is an activity or a practice, equivalent to prostitution or masturbation that may become a habit. It is neither genetic nor hereditary.**

Customs or habits are alcoholism, drug addiction, smoking, also, including, positive practices such as carpentry, masonry, engineering, taxi driving, activities may be changed without interpreting such change as a modification of the genetic code of the people practicing same.

Customs, habits and abilities are conducts learnt or voluntarily acquired by people and repeated by routine until they become mechanic. They are acquired features and not inherited genetically.

Therefore, if homosexuals stop practicing said habit, they are not homosexuals. Alcoholics that stop drinking may not be called drunks anymore. If a prostitute abandons the profession, she stops being a prostitute. If a pilot stops flying and becomes an entrepreneur and lives his previous trade, she should no longer be called a pilot.

From the previous I can conclude that individuals that regenerate from a habit or custom are, for that fact, reintegrated and cannot and should not be distinguished for the habit they used to practice.

In other words, it is a proven fact that the genetic code of human beings has an X chromosome that is located at what is called Pair 23. When Pair 23 is XX the gender of the individual is chromosomally called female. In the event the pair is XY, the sex of the individual shall be chromosomally called male.

This means that in the genetic code regulating the human species —which is an invariable and eternal law of Nature— only two genders exist: male and female: there is no homosexual chromosome, therefore, there is no "third gender."

The sexual capability to reproduce is in fact genetically inherited; it is a law of Nature because it is universal in all humans, but the sexual act is voluntary, it is not a law of Nature, it is not obligatory. Each individual may voluntarily and consciously decide when and how to direct his sexual impulses.

Hermaphroditism appearing in some vegetable and animal species has as a characteristic a mixed reproduction system in the same individual that is capable of producing both male and female gametes. This quality is identical for all individuals of the same species (animal or vegetable) and is transmitted by universal inheritance.

But in the human species this condition may not happen because one same individual male or female may not produce sperm and eggs at the same time, in any case it would be a pathology by the most important thing is that said condition is not inherited; therefore, it may not be considered as a law of Nature.

I think that homosexuality is not a question of identity but an addiction that may be changed with new conducts. It not hereditary either; there are no genes regulating or controlling the manner to satisfy sexual pleasure. The assertion made by medical doctors and scientists regarding the genetic inheritance of homosexuality is a lie.

When homosexuals refer to being born homosexual they are referring to their condition of being born a male with female features.

These features are in fact congential and constitute a scientific fact but they are being confused by assuming that their sexual inclination when they decide the course of their sexuality shall make them homosexual.

Their decision to practice homosexuality is strictly an act of will and may be disregarded.

Nether is homosexuality an "orientation" because homosexuals are not directed or guided to practice.

In some serious scientific media (Whithead Institute of Massachusetts) investigations have been done that have directed them to conclude that homosexuality is genetically hereditary.

Those investigations have not had sufficient statistical support due to the lack of evidence to be asserted, and they are, therefore, preliminary investigations about the problem taking into consideration the arguments shown before, regarding the genetic code in the human beings as invariable, I can assert, without any doubt, that said investigations are prone to fail.

Homosexuality has the only objective to satisfy the sexual pleasure of the individual in this manner, bestiality has the same goal and it has the same level of sexual satisfier.

Masturbation, even though it is a self-satisfier, has also the same objective: to satisfy the sexual pleasure.

By failing to comply with the law of Nature, which is reproduction, these practices are unnatural.

At the end of the 1970s an illness was detected with unknown characteristics that had no cure. The majority of the people that were sick belonged to the homosexual community, intravenous drug users and blood transfusion recipients.

It was in the city of San Francisco and New York where an alarming incidence of people infected with this virus later known as AIDS (acquired immune deficiency syndrome) also known as HIV-aids.

Probably, this viral infection existed since the origins of the Human Race but was undetected for lack of scientific resources. Scientific investigators managed to isolate the aids virus at the beginning of the 1980s.

At the beginning, scientists thought that the virus was contracted in Africa by practicing bestial sex and that a green monkey had infected some individuals and these in turn infected other humans by having sexual relations. This first assumption about the origin of the HIV virus was eliminated when they tried to infect the monkeys with the AIDS virus and said monkeys were already immune.

From the foregoing it can be deducted that this illness is a venereal disease, like syphilis, gonorrhea, and others, and it originated from the sexual relations between humans. It is proven that it is not transmitted by heterosexual intercourse.

Aids is not caused by syphilis, gonorrhea and other illnesses that are transmitted by heterosexual relations, therefore, I conclude that aids is a product of homosexual relations.

The work of scientists now shall be to investigate the conditions in which the illness originates and lay the basis to prevent it, as well as to provide the media and curative methods and look for the eradication of such illness.

I think that individuals should be instructed on the consequences of practicing homosexuality, as if it propogates, it could exterminate the human species for lack of reproduction.

Chapter VIII

Human Rights

We accept as evident truth that all human beings are created equal before the laws of Nature and that the Creator has granted certain inalienable rights that are, among others: life, freedom and the pursuit of happiness.

All living things have been created to live. This is the most important commandment of the laws of Nature governing all living creatures.

In order to fulfill this purpose, all organisms must comply with these obligations imposed by Nature, such as to breathe, eat, drink, sleep, eliminate waste, etc.

Human beings, besides the entire foregoing, have the capacity to preserve their health, look for comfort, live in society and with dignity and achieve longevity.

Then, nobody must impede a human being from fulfilling his obligations and in turn, said human being is obligated to respect the same obligations of survival of the rest of the human beings. This is from where the so-called "Right to Life" comes.

Because we have been created equal in front of the laws of Nature, all humans have the same rights.

Do you consider correct that some governments impose the death penalty to a member of society accused of a premeditated homicide, kidnaping or rape?

World governments are constituted by humans of the same nature of their constituents and based upon the principle that no human is superior, to another I am able to conclude that no one has the right to deprive any of their peers of their life, but they may indeed deprive said person of their freedom as a penalty for said person's transgression.

Dear reader: I understand that there are occasions where a person is the victim of a serious crime, like rape, kidnaping or unjustified injuries and that in such moment passions negative like hate, resentment, anger or vengeance may surface to redress the offense received, but, the human being is obligated to control such passions in order to achieve inner peace and have emotional stability.

Even though it may be difficult to accept in dark moment, forgiveness is the best way to achieve inner peace because peace is the foundation of happiness.

In other words, for Nature, all individuals are the same and have the same right to use its goods to fulfill their obligations to live, enjoy and reach the most longevity.

Nature belongs to everyone and is for everyone: the sun shines for all, we breathe the same air, the rain falls on everyone, seawater, rivers and snow belong to us all.

If we deprive a human being of his right to produce food, from cultivating the land, from water to drink or to irrigate the land, etc., said individual shall not be able to achieve the most longevity and his survival will be in danger and all human beings are jointly responsible of the violation of human rights.

In the same manner I think that all human beings have the same right to go to other places in the world to look for nourishment and fulfill their longevity and survival commandment.

Planet Earth is just one; it has no divisions. Countries are an invention of human beings that divided the Earth, separating it, and thereby failing to comply with the laws commanded by Nature.

We humans are designed to live in society and to mix with other human beings without the distinction of race, religion or social or economic status.

Human migration around the globe to look for nourishment is a Universal Right of all individuals that must be respected by all human societies and must be adequately regulated by the political authorities.

Representatives of human societies must dictate the legal precepts to guarantee every one of their members the free exercise to fulfill their obligations and duties demanded by the laws of Nature: survival, conservation of the species and longevity.

These obligations are incorporated into the genetic code of each and every one of the human beings –that is why they are universal and are converted into rights for all humans to be able to enforce them before their peers, and of course before all the authorities of the world.

First comes duty and then comes right. Rights come from the obligation that all humans have to respect one another.

In fact, these rights are recognized in the Universal Declaration of Human Rights approved in the city of Paris in the year 1948 for all countries member of the General Assembly of the United Nations.

The purpose of the so-called Human Rights is that all individuals reach a longevity of one hundred and twenty years without the distinction of race, sex, nationality or religion because Nature makes no distinction between one and another human being.

Human rights originate, without exception, from the obligations and duties directed by the laws of Nature.

I deem prudent to ask my reader the following question: do you think that socialism is good as a social and economic organizational system?

The laws of Nature establish that no human being should abuse another. This is the reason why I consider that the rational humanist principles (ill-called socialism) are correct but unfortunately it is impossible to plant them at the present because human societies are poorly constructed because they constantly contravene these laws.

The foregoing is proven by the inequality existing between the rich and the poor. The laws of Nature do not prevent the presence of rich and poor people in the world. What they command is that No be people should be unable to comply with the commandment to live one hundred and twenty years, and there are millions of people dying of hunger not being able to live half of the longevity scheduled for them by the laws of Nature. While there is one hungry individual in the world, there should not be one rich one.

Human beings must do whatever is necessary to keep alive. This is an obligation, a commandment. It is not a right before anybody else, is a duty to oneself, to the Nature of the human being and not to others.

These affirmations are also founded in the dependency shown by all human beings on Nature.

To the laws of Nature, humans have no liberties or rights, only obligations. Once the obligations commanded by Nature are fully satisfied they become rights to be enforced by the individual before others, and the governments of the world should respect those rights. If human beings do not respect said laws, they are destroyed. There is no alternative.

World rulers know that they have the obligation to protect the life of their constituents providing justice, education, safety, peace, medicines, jobs and rest in order to live with dignity.

To achieve those goals, the state dictates laws to protect the execution of those natural obligations and penalize those who prevent others from complying with them.

Objective law is the regulation of the obligations to be met by all human beings, and it covers in justice the actions that occur between humans: Labor, Criminal, Land, and Civil Law among others. Objective Law is based upon Natural Law.

On the other hand, I think that the definition of dignity given by some dictionaries is an ambiguous one.

In my opinion, dignity is the right all human beings have to be respected in their fulfilment of

their own obligations such as life, nourishment, housing, rest, work, reproduction, and health.

Said obligations are known as Natural Law because Nature is the only one who has active participation.

I consider that <u>dignity is the origin of the so-called Human Rights</u> and it is a mistake to name an obligation a right. This obligation is the one giving rise to the right of men.

When a human being fails to comply with the duties imposed by dignity, he becomes undignified as does one who fails to respect the duties of their peers.

Compliance with those obligations give a human being the right to be respected by their peers. Each human being, when respecting and helping their peers to comply with their duties is acting ethically.

To live is an obligation, not a duty to anybody, but an obligation to oneself and the individual only is accountable for his own behavior internally, to one's own conscience.

The right, the juridical order established by each human society (or country) is not universal, because it governs only the external behavior of human beings; therefore, such right is not a law of Nature.

In fact, this juridical order is not equal and does not obey the same rules in all individuals of the different human societies. On the other hand, the laws of Nature rule the internal and genetic conduct of all human beings that is why it is universal.

"All men have been created equal." Here the term men are generic; it refers implicitly to women also. This principle set forth since the seventeenth century is accepted to this date without any difference of opinion because it has been proven experimentally, this is, scientifically.

The principles of Natural Law are based on the human Nature; they are inscribed in the genetic code of the individuals since their creation and that is why they are enforced worldwide.

Human Rights are an expression of the Natural Law and the Universal Declaration of Human Rights is based upon the laws of Nature, specifically in the laws of morality giving rise to the Universal Right or Natural Right.

Right refers to the collective values while morality pursues the realization of personal values.

No organization or governmental entity has the authority to give human individuals rights not coming from the laws of Nature.

Karl Larenz[52] says that in his traditional meaning, the term Natural Law has meant always an immutable

---

[52] Karl Larenz.- *Metodología de la Ciencia del Derecho.-* Editorial Ariel.- España.- 1979.

right, at least in its fundamental lines and that it is valid equally for all nations of the world because it is based in the essence of human beings.

Articles first and third of the Universal Declaration of Human Rights literally read as follows: *"All humans are born free and equal in dignity and rights..."* and *"Everyone has the right to life, liberty and security of person."*

In my opinion, the thirty articles contained in the Universal Declaration of Human Rights may be summarized in one, "The survival and longevity of a human being is the obligation and responsibility of each and every one of his and her peers."

Opulent countries should not take advantage of the ones that live in misery nor take or limit their right to the fruits of the Earth, leaving them without food or without the opportunity to work the land or without a place to live because they would be preventing the accomplishment of this universal law.

Powerful nations are the ones that have created poverty by exploiting people and abusing the natural resources of the underdeveloped countries, and they have the obligation to help said countries improve their life conditions to comply also with the commandments of the laws of Nature.

Developed countries and the privileged social groups that have forgotten or disregarded compliance with the laws of Nature are tilling their own extinction.

Countries living in misery shall keep a constant and unavoidable fight to vindicate this law, and finally they shall fight to survive. This confrontation shall conclude with the extinction of human society.

In order to survive, human beings must renounce their selfishness to benefit society. Serving others is showing love for the human species, a law of Nature, and therefore obligatory.

If society does not comply with the commandments of the laws of Nature to end poverty, I can predict that in an indefinite period (eon), the human species will have massively destroyed itself and civilization will have to begin from cero.

I can also speculate that it will take civilization hundreds of thousands of years to reach the levels we know now, but the laws of Nature will not change, they will be present in the entire course of the new civilization and until de end of the new development in the cycle of the human species.

Guessing, I can also affirm that the new civilization, up to and through the end of that new cycle of the human species, will be also destroyed if it does not comply with the laws of Nature.

Conclusion

Within the laws of Nature is the only parameter we can take as infallible and universal to avoid making mistakes in judgment. We can consider these laws as the sole and absolute truth human knowledge is able to prove.

Knowing of the laws of Nature is indispensable and necessary to correct the erratic conduct of humans because men and women are a part of Nature, and they are obligated to live in harmony with these laws that are eternal and invariable and applicable for all human individuals without exception of race, creed, age or social status.

All, absolutely all the Universe, including thoughts, feelings, emotions, conducts and the activities of human beings, are subject to these laws.

All facts and phenomena of Nature serve a program pre-established by the same laws of Nature. Every occurring event has a previous cause. There is no chance or fate in Nature, but an order in every part of the Universe.

Plants and animals live according to their existence and in the place assigned to them by Nature following the program already incorporated into their genetic code. They live and feed without harming the environment because they lack the capacity to breach the rules of Nature.

But human beings have the capacity to act against these laws and that has caused the serious problems that humanity is suffering now.

Sex abuse has been one of the main causes of problems in humanity because it has caused the overpopulation of the Planet, and at the same time, it has originated water and food shortage, poverty, deforestation, the pollution of rivers, lakes, seas and the environment as well as many other problems of the human race.

The laws of Nature may be discovered by Science. Science is the only means that humans have to find truth and discover the benefits provided by applying these laws to produce goods to aid humans to survive, find comfort and reach longevity.

The main law of Nature for all living creatures is survival and all living organisms are obligated to obey theses specific laws that are already engraved in the genetic code of each species. Within these laws, we can find the ones that govern processes such as digestion, metabolism, breathing, waste elimination, etc.

Besides the features mentioned above, which are common to living creatures, human beings have specific roles that differentiate them from the other species.

Within said roles are, among others, are the capacity to discern between good and bad, the capacity to produce thoughts, the instinct towards what is sacred or religious, the capacity to self-reflect, reflect over one's own sensations, love, and wanting to develop an idea, etc.

Selfishness is the antithesis of love, and it is one of the great evils of humanity provoking disorder in human societies.

Negative feelings of the human being, such as anxiety, anxiousness, obsession, nervousness, vanity, selfishness, hate, stress, resentment, ambition, vengeance, envy, lust, greed, anger, rage, jealousness, sadness, and others, cause great damage in the human body as well as the damage caused to society.

Actions arising from these negative feelings destroy the possibility to obtain equilibrium for which the human being was created.

Human instincts exist for the individuals to comply with the commandments imposed by the laws of Nature. These laws compel humans to eat, drink, sleep, reproduce, etc.

In order to comply with these obligations, instincts make humans feel hunger, thirst, sleep, sexual desire, etc., and the human individual is programmed to feel satisfaction after following these instincts, but exceeding their satisfaction is the cause of obesity, alcoholism, indolence, lust, overpopulation, and as a consequence of overpopulation, poverty, famine, etc.

All goods and fruits of the Earth belong equally to every individual whether human or not. Everyone has the same right to those goods because they are necessary for their survival and longevity; human beings are responsible for their own survival and longevity and of the rest of the living creatures.

The so-called "Human Rights" created by the civil societies are created so the individual is able to meet the commandments of the laws of Nature. These "rights" can be enforced before the rest of the individuals so no one prevents another from meeting their primary obligations of keeping alive and seeking longevity.

From the foregoing we may conclude that failure to comply with any of the laws of Nature is the cause of all illnesses, depression, suffering and pain in humans.

The analysis and comprehension of these concepts was precisely the purpose of this book, and the wish of the author is that we be better humans that can live in health, prosperity and inner peace to achieve happiness.

Bibliografía

- Sócrates.- *Diálogos de Platón.*- Digital edition: http://vagabundeoresplandeciente.wordpress.com/2008/04/10/antigona-de-sofocles-y-la-distincion-juridica-cardinal/

- Marco Tulio Cicerón.- *De Republica.*- Digital edition: http://www.eleutheria.ufm.edu/Articulos/060316_CAPITULO_II_DE_LA_BUSQUEDA_DE_LA_COMUNIDAD.htm

- Marco Aurelio.- *Meditaciones.*- Tomado de: *El Humanismo Jurídico en San Marcos.*- José Antonio Ñique de la Peña.- Digital version: http://sisbib.unmsm.edu.pe/bibvirtualdata/tesis/human/%C3%91ique_pj/pdf/parte1.pdf

- Cleantes, *Himno a Zeus.*- "Fr. Stoic." I, 527 - cf. 537). Digital version: http://ec.aciprensa.com/l/logos.htm

- Filón de Alejandría.- *El heredero de los bienes divinos "Quis rerum divin. haeres sit"*. Digital citation: http://mb-soft.com/believe/tsxtm/logos.htm

- Tertuliano.- *Adv. Hermogenem.*- 44.- Digital edition: http://ec.aciprensa.com/l/logos.htm

- The King James Version of the Holy Bible.- Digital Edition: www.davince.com/bible
http://d23a3s511qjyz.cloudfront.net/wp-content/uploads/2012/09/King-James-Bible-KJV-Bible-PDF.pdf

- Ashley Montagu.- *¿Qué es el Hombre?* (2ª. Ed).- Editorial Paidós Ibérica.- 1987.

- Paul Davis.- *Dios y la Nueva Física.*- Salvat Editores, S.A.- Barcelona.- 1994.

- El alma en Aristóteles.- Digital version: http://filosofia.laguia2000.com/filosofia-griega/el-alma-en-aristoteles

- Filosofía Griega.- Origen de la Filosofía-Presocráticos-Sofistas y Sócrates.- Digital citation: http://www.e-torredebabel.com/Historia-de-la-filosofia/Filosofiagriega/Presocraticos/Alma.htm

- Albert Eintein.- Citas.- Digital vertion: http://es.wikiquote.org/wiki/Albert_Einstein.

- Marcelino Cereijido Mattioli.- *Por qué no Tenemos Ciencia.*- Siglo xxi Editores, S.A. de C.V.- México, 2004.

- Marcelino Cereijido.- *Ciencia Sin Seso.*- Siglo xxi Editores, S.A. de C.V.- sexta edición.- México, 2005.

- Baruch de Spinoza.- *Ética demostrada según el orden geométrico.*- Editorial Orbis S.A.- Barcelona, España.- 1980.

- Marvin Harris.- *El Desarrollo de la Teoría Antropológica.*- Siglo XXI de España Editores.- 2008.- Pág. 16.

- Elizardo Martínez Vergara.- *La Doctrina del Derecho Natural de Hugo Grocio.*- Universidad San Francisco De Asís.- La Paz, Bolivia.- 2006.

- Miguel Angel Cuevas Guinto.- *Los Países, Leyes y Constituciones.*- Digital vertion: http://paises-y-leyes.blogspot.mx/2012_04_29_archive.html

- John Locke.- *Dos tratados del gobierno civil (1690).*- Fragmentos citados en *Textos fundamentales para la historia.*- Alianza Editorial.- Madrid, España.- 1982.- Citado en Los Países, Leyes y Constituciones.- Digital edition: http://paises-y-leyes.blogspot.mx/2012_04_29_archive.html

- Santo Tomás.- *La Suma Teológica*.- 1-2, 100, 8, ad2.- Digital citation: http://es.scribd.com/doc/18678110/resumen-santo-tomas-de-aquino-suma-teologica.

- Regis Jolivet.- *Tratado de Filosofía. Moral.*- Editorial Lohle.- Buenos Aires, Argentina.- 1959.

- Interview by Larry King of Stephen Hawking, December 25, 1999 at 9:00 pm, on CNN television: http://www.psyclops.com/hawking/resources/cnn.html

- Leonardo Da Vinci.- *Escritos Sobre La Ciencia.*- Digital version: http://www.slideshare.net/rurenagarcia/escritos-de-leonardo-sobre-ciencia

- F. David Peat.- *Sincronicidad.- Puente entre mente y materia.*- Digital citation: http://alberkrip.files.wordpress.com/2008/06/peat-f-david-sincronicidad-puente-entre-mente-y-materia.pdf

- Rafael De Pina y Rafael De Pina Vara.- *Diccionario de Derecho.*- 32ª. Edición.- Editorial Porrúa.- México.- 2003.

- Jaime Carder.- *La Naturaleza Habla.*- Editorial CLIE. Barcelona. 1973.

- Terencio Arteaga Pincay.- *Conferencia dictada el sábado 8 de julio del 2006* por: Dr. Enrique mármol Palacios y Dra. Marlene Sánchez.- Facultad de Jurisprudencia, Ciencias Sociales y Políticas.- Universidad de Guayaquil.- Ecuador.-

- René Descartes y el Legado del Dualismo Mente-Cuerpo.- Digital Citation: http://platea.pntic.mec.es/~macruz/mente/descartes/descartes.html

- Francisco J. Ayala.- *La Naturaleza Inacabada.-* Salvat Editores.- 1994.

- Esther Jacob.- *Mi Cuerpo.-* editorial: Secretaría de Educación Pública / Conaculta.- México.- 2000.

- Real Academia Española.- *Diccionario de la Lengua Española.-* Editorial Espasa Calpe, S.A.- Madrid, España.- 1992.

- Karl Larenz.- *Metodología de la Ciencia del Derecho.-* Editorial Ariel.- España.- 1979.

- Wikipedia – The free encyclopedia.- Digital Version: http://en.wikipedia.org/

- Juan Pablo II.- Encíclica *Dives in misericordia.-* Digital version: http://www.instituto-social-leonxiii.org/index.php/estudios/844-la-animacion-de-la-caridad-en-la-comunidad

- Liz Hodgkinson.- *El Sexo No Es Obligatorio.-* Javier Vergara Editor, S.A.- España.- 1988.

- The Book of the year 1988.- *Encyclopædia Britannica.-* Ed. Encyclopædia Britannica, Inc.- U.S.A.- 1989.

- The Book of the year 1998.- *Encyclopædia Britannica.-* Ed. Encyclopædia Britannica, Inc.- U.S.A.- 1999.

- Un Statistics Division, Department of Economic and Social Affairs.- *"World Population Prospects: The 2008 Revision".-* Digital version: http://www.geohive.com/earth/pop_gender.aspx

- George Berkeley.- *Tratado sobre los Principios del Conocimiento Humano.-* Digital version: http://es.wikipedia.org/wiki/Tratado_sobre_los_principios_del_conocimiento_humano

- Enciclopedia Microsoft. *Encarta.* 2002.- 1993-2001.- Microsoft Corporation.

- André Lalande.- Vocabulario Técnico y Crítico de la Filosofía.- Librería El Ateneo.- Buenos Aires 1953.

- Phrase de David Hume.- Digital Citation: http://akifrases.com/frase/115825

www.ingramcontent.com/pod-product-compliance
Lightning Source LLC
Chambersburg PA
CBHW032020170526
45157CB00002B/782